IET POWER AND ENERGY SERIES 77

Wide-Area Monitoring of Interconnected Power Systems

Other volumes in this series:

Volume 1 **Power circuit breaker theory and design** C.H. Flurscheim (Editor)
Volume 4 **Industrial microwave heating** A.C. Metaxas and R.J. Meredith
Volume 7 **Insulators for high voltages** J.S.T. Looms
Volume 8 **Variable frequency AC motor drive systems** D. Finney
Volume 10 **SF_6 switchgear** H.M. Ryan and G.R. Jones
Volume 11 **Conduction and induction heating** E.J. Davies
Volume 13 **Statistical techniques for high voltage engineering** W. Hauschild and W. Mosch
Volume 14 **Uninterruptible power supplies** J. Platts and J.D. St Aubyn (Editors)
Volume 15 **Digital protection for power systems** A.T. Johns and S.K. Salman
Volume 16 **Electricity economics and planning** T.W. Berrie
Volume 18 **Vacuum switchgear** A. Greenwood
Volume 19 **Electrical safety: a guide to causes and prevention of hazards** J. Maxwell Adams
Volume 21 **Electricity distribution network design, 2nd edition** E. Lakervi and E.J. Holmes
Volume 22 **Artificial intelligence techniques in power systems** K. Warwick, A.O. Ekwue and R. Aggarwal (Editors)
Volume 24 **Power system commissioning and maintenance practice** K. Harker
Volume 25 **Engineers' handbook of industrial microwave heating** R.J. Meredith
Volume 26 **Small electric motors** H. Moczala *et al.*
Volume 27 **AC-DC power system analysis** J. Arrillaga and B.C. Smith
Volume 29 **High voltage direct current transmission, 2nd edition** J. Arrillaga
Volume 30 **Flexible AC transmission systems (FACTS)** Y-H. Song (Editor)
Volume 31 **Embedded generation** N. Jenkins *et al.*
Volume 32 **High voltage engineering and testing, 2nd edition** H.M. Ryan (Editor)
Volume 33 **Overvoltage protection of low-voltage systems, revised edition** P. Hasse
Volume 36 **Voltage quality in electrical power systems** J. Schlabbach *et al.*
Volume 37 **Electrical steels for rotating machines** P. Beckley
Volume 38 **The electric car: development and future of battery, hybrid and fuel-cell cars** M. Westbrook
Volume 39 **Power systems electromagnetic transients simulation** J. Arrillaga and N. Watson
Volume 40 **Advances in high voltage engineering** M. Haddad and D. Warne
Volume 41 **Electrical operation of electrostatic precipitators** K. Parker
Volume 43 **Thermal power plant simulation and control** D. Flynn
Volume 44 **Economic evaluation of projects in the electricity supply industry** H. Khatib
Volume 45 **Propulsion systems for hybrid vehicles** J. Miller
Volume 46 **Distribution switchgear** S. Stewart
Volume 47 **Protection of electricity distribution networks, 2nd edition** J. Gers and E. Holmes
Volume 48 **Wood pole overhead lines** B. Wareing
Volume 49 **Electric fuses, 3rd edition** A. Wright and G. Newbery
Volume 50 **Wind power integration: connection and system operational aspects** B. Fox *et al.*
Volume 51 **Short circuit currents** J. Schlabbach
Volume 52 **Nuclear power** J. Wood
Volume 53 **Condition assessment of high voltage insulation in power system equipment** R.E. James and Q. Su
Volume 55 **Local energy: distributed generation of heat and power** J. Wood
Volume 56 **Condition monitoring of rotating electrical machines** P. Tavner, L. Ran, J. Penman and H. Sedding
Volume 57 **The control techniques drives and controls handbook, 2nd edition** B. Drury
Volume 58 **Lightning protection** V. Cooray (Editor)
Volume 59 **Ultracapacitor applications** J.M. Miller
Volume 62 **Lightning electromagnetics** V. Cooray
Volume 63 **Energy storage for power systems, 2nd edition** A. Ter-Gazarian
Volume 65 **Protection of electricity distribution networks, 3rd edition** J. Gers
Volume 66 **High voltage engineering testing, 3rd edition** H. Ryan (Editor)
Volume 67 **Multicore simulation of power system transients** F.M. Uriate
Volume 68 **Distribution system analysis and automation** J. Gers
Volume 69 **The lightening flash, 2nd edition** V. Cooray (Editor)
Volume 70 **Economic evaluation of projects in the electricity supply industry, 3rd edition** H. Khatib
Volume 78 **Numerical analysis of power system transients and dynamics** A. Ametani (Editor)
Volume 905 **Power system protection, 4 volumes**

Wide-Area Monitoring of Interconnected Power Systems

Arturo Román Messina

The Institution of Engineering and Technology

Published by The Institution of Engineering and Technology, London, United Kingdom

The Institution of Engineering and Technology is registered as a Charity in England & Wales (no. 211014) and Scotland (no. SC038698).

First published 2015

The Institution of Engineering and Technology
Michael Faraday House
Six Hills Way, Stevenage
Herts, SG1 2AY, United Kingdom

www.theiet.org

British Library Cataloguing in Publication Data
A catalogue record for this product is available from the British Library

ISBN 978-1-84919-853-0 (hardback)
ISBN 978-1-84919-854-7 (PDF)

Typeset in India by MPS Limited
Printed in the UK by CPI Group (UK) Ltd, Croydon

Contents

Preface ix

1 Wide-area monitoring and analysis systems **1**
1.1 Introduction 1
1.2 Wide-area monitoring systems: a conceptual overview 2
1.3 Data collection and management 3
1.4 Challenges of future smart monitoring and analysis systems 4
References 5

2 Wide-area monitoring system architectures **9**
2.1 Introduction 9
2.2 WAMS architectures 9
2.2.1 Centralized WAMS architectures 11
2.2.2 Hierarchical WAMS architectures 12
2.2.3 Hybrid WAMS architectures 13
2.3 Issues in data fusion 13
2.3.1 Data 13
2.3.2 Intelligent synchrophasor data fusion 14
2.3.3 Power system data fusion strategies 16
2.3.4 General framework for data assimilation 20
2.4 Relationship between multiblock and single-block models 22
References 23

3 Spatio-temporal modeling of power system dynamic processes **27**
3.1 Introduction 27
3.2 Visualization of large space-time measurement data 28
3.3 Spatio-temporal modeling of multivariate processes 29
3.3.1 Empirical orthogonal function (EOF) analysis 29
3.3.2 SVD-based proper orthogonal decomposition 33
3.3.3 Departure from mean value 36
3.4 Spatio-temporal interpolation methods 37
3.4.1 Background 37
3.4.2 Similarity measures 38
3.4.3 Spatial structures 40
3.4.4 Derivation of weights 40
3.4.5 Practical issues 41

3.5 Dimensionality reduction 42
3.5.1 Proximity (similarity) measures 42
3.5.2 Nonlinear spectral dimensionality reduction 43
3.6 Motivational example 47
3.6.1 Small-signal response 48
3.6.2 Large system response 48
3.6.3 Statistical analysis 49
3.7 Sensor placement 51
3.7.1 Problem formulation 53
3.7.2 Constrained sensor placement 54
References 58

4 Advanced data processing and feature extraction 63
4.1 Introduction 63
4.2 Power oscillation monitoring 64
4.3 Time-frequency representations 65
4.3.1 Hilbert–Huang analysis 65
4.3.2 Wavelet analysis 72
4.3.3 The Teager–Kaiser operator 75
4.3.4 Dynamic harmonic regression 76
4.4 Mutivariate multiscale analysis 81
4.4.1 Multi-signal Prony analysis 82
4.4.2 Koopman analysis 83
4.5 Response under ambient stimulus 87
4.5.1 Formulation of the model 87
4.5.2 Modal response 89
4.5.3 Ensemble system response 90
4.6 Application to measured data 90
4.6.1 HHT analysis 92
4.6.2 Wavelet analysis 94
References 96

5 Multisensor multitemporal data fusion 101
5.1 Introduction 101
5.2 Data fusion principles 101
5.3 Data pre-processing and transformation 104
5.3.1 Bandpass filtering and denoising 104
5.3.2 Local-level fusion 105
5.4 Feature extraction and feature selection 105
5.4.1 Feature extraction 105
5.4.2 Data compression 106
5.4.3 Individual scales 109
5.4.4 Filtering and multiscale monitoring 109
5.5 Multisensor fusion methodologies for system monitoring 111
5.5.1 Single-scale analysis 112
5.5.2 Nonlinear PCA using auto-associative neural networks 112

5.5.3 Multiblock POD (PCA) analysis 113
5.5.4 Nonlinear PCA 119
5.5.5 Blind source separation 119
5.6 Other approaches to multisensor data fusion 124
References 128

6 Monitoring the status of the system 131
6.1 Introduction 131
6.2 Power system health monitoring 132
6.3 Disturbance and anomaly detection 132
6.4 Modal-based health monitoring methods 134
6.4.1 Filtering and data conditioning 134
6.4.2 Entropy and energy 138
6.4.3 Entropy-based detection of system changes 141
6.5 Wide-area inter-area oscillation monitoring 143
6.5.1 Case A 143
6.5.2 Case B 145
6.6 High-dimensional pattern recognition-based monitoring 146
6.6.1 Sparse diffusion implementation 146
6.6.2 Data clustering 148
6.6.3 Numerical example 148
6.6.4 Hybrid schemes 150
6.7 Voltage and reactive power monitoring 150
6.7.1 Measured data 150
6.7.2 Statistical approach to voltage monitoring 151
6.7.3 Complex POD/PCA analysis 155
References 156

7 Near real-time analysis and monitoring 159
7.1 Introduction 159
7.2 Toward near real-time monitoring of system behavior 159
7.3 Data processing and conditioning 160
7.3.1 Wavelet denoising and filtering 160
7.3.2 EMD-based filtering 162
7.4 Damage detection from changes in system behavior 163
7.4.1 Event trigger 164
7.4.2 Event detection based on linear filtering 164
7.4.3 An illustration 166
7.5 Time-series approaches to detection of abnormal operation 166
7.5.1 Near real-time implementations 166
7.5.2 Near real-time implementation of the Hilbert transform 169
7.5.3 Local mean speed 173
7.6 Pattern recognition-based disturbance detection 176
7.7 Sliding window-based methods 177
7.7.1 Local HHT analysis 177

7.7.2 Numerical example 180
7.7.3 Sliding window-based Koopman mode analysis 181
7.8 Recursive processing methods 182
7.8.1 State-space model for linear regression 182
7.8.2 Adaptive tracking of system oscillatory modes 183
References 188

8 Interpretation and visualization of wide-area PMU measurements 191
8.1 Introduction 191
8.2 Loss of generation oscillation event 191
8.2.1 Operational context 192
8.2.2 Recorded measurements 192
8.3 Analysis and visualization of recorded data 196
8.3.1 Mode shape characterization 196
8.3.2 Damping estimation 197
8.3.3 Instantaneous parameters 197
8.3.4 Multitemporal, multiscale analysis of measured data 205
8.3.5 Performance evaluation 209
8.4 Pattern recognition analysis 211
8.4.1 Diffusion map analysis 211
8.4.2 Comparison with other approaches 214
8.5 POD/BSS analysis 216
8.6 Validation of power system model 218
8.6.1 Small signal performance 218
8.6.2 Large system performance 218
8.7 Evaluation of control performance 221
References 225

Appendix A Physical meaning of proper orthogonal modes 227

Appendix B Data for the five-machine test system 231

Appendix C Masking techniques to improve empirical mode decomposition 235

Index 237

Preface

In the last few years, the interest in monitoring and analysis of key system variables throughout the transmission and distribution system has significantly increased due to the need to assess the power system health in near real-time. The emergence of new sensors, advanced communication systems, and improved processing techniques makes real-time system-wide monitoring increasingly possible. These advances result in large data sets that must be integrated to reduce uncertainty in power system security and reliability assessment.

The purpose of this book is to provide a comprehensive treatment of advanced data-driven signal processing techniques for the analysis and characterization of both ambient power system data and transient oscillations resulting from major disturbance. Inspired by recent developments in multisensor data fusion, multitemporal data assimilation techniques for power system monitoring are proposed and tested in the context of modern wide-area monitoring system (WAMS) architectures. Recent advances in understanding and modeling nonlinear, time-varying power system processes are reviewed and factors affecting the performance of these techniques are discussed.

A number of algorithms and examples are presented throughout the text as an aid to understanding the basic material provided. Challenges involved in realistic monitoring, visualization, and analysis of actual disturbance events are emphasized and examples of applications to a wide range of power networks are provided.

Structurally, the book is divided into three basic parts. The principal theoretical thrust of the book is embodied in Chapters 3 through 7. Chapters 1 and 2 examine the most important problems in WAMS, stressing the problem of data management and data fusion. They also introduce simulation methodology. Chapters 3 and 4 discuss the development of advanced algorithms for the analysis and characterization of spatio-temporal dynamics and illustrate and compare some of the proposed analytical procedures.

Chapters 5 through 7 examine the application of novel multivariate, multitemporal data analysis techniques to the analysis and visualization of synchrophasor data. Examples are used throughout to demonstrate the application of the theory. Chapter 8 examines the application of WAMS strategies to monitor and visualize multiple streams of phasor measurement unit (PMU) data.

The appendixes at the end of the book provide necessary complementary information to the book.

Arturo Román Messina
March 2015

Chapter 1

Wide-area monitoring and analysis systems

1.1 Introduction

The development of advanced wide-area monitoring systems (WAMS) based on synchrophasor technology provides unprecedented views of power system dynamic behavior with increased resolution and accuracy [1, 2]. In addition to the growth in the amount of data, the variety of measurement devices has also increased. In this context, advances in the development and installation of inexpensive, low-voltage recording devices have resulted in the deployment of a large number of sensors that transmit data to specialized data concentrators. As the size and complexity of power grids continue to increase, real-time monitoring and forecasting of dynamic processes become increasingly important. This increase in both the volume and variety of the data requires advances in methodology to automatically understand, process, and summarize the data.

Fast, high-quality synchrophasor measurements of voltage and current phasors using signals from a global positioning satellite have the potential to greatly enhance wide-area visibility and result in enhanced system security and reliability [3–5].

Advanced applications in wide-area monitoring encompass the implementation of situational awareness systems including disturbance alert, event location triangulation and oscillation detection, early warning systems, power system oscillatory monitoring, and other advanced features [6]. At the core of these systems are intelligent sensing and signal processing and communication techniques to make optimal use of wide-area data.

Analysis and characterization of time-synchronized system measurements require smart monitoring tools that are adaptable to the varying system conditions, accurate and fast, while reducing the complexity of the data to make them comprehensible and useful for automated control and real-time decision.

This chapter provides an overview of key principles in wide-area monitoring architectures. Models, applications, and areas of improvement in real-time system monitoring and key research directions in the area of data management are described and highlighted.

1.2 Wide-area monitoring systems: a conceptual overview

Over the last two decades, various forms of WAMS have been developed to monitor power system oscillatory behavior. At its core, a wide-area monitoring system is an intelligent, continuous identification system of power system status.

The key components of a WAMS are shown in Figure 1.1. Conceptually, the WAMS consists of different components such as frequency disturbance recorders, phasor measurement units (PMUs), digital relays, and advanced communication links and signal processing techniques. Synchronized measurement technology provides phasor data at rates up to 60 samples per second which allows near-real-time monitoring of system behavior.

The WAMS structure is hierarchical and can be split into two major levels; regional and global. At a local level, sensed information is automatically collected, synchronized, and archived by a monitoring and control center known as Phasor Data Concentrator (PDC) [7, 8]. This information is then sent to a global data concentrator for real-time dynamics monitoring (RTDM) system, wide-area control (WAC), and wide-area protection (WAP) for inference, estimation, and control and protection purposes.

The input data may be imperfect, correlated, dynamically inconsistent, and in disparate forms or modalities. Monitoring provides critical data to be processed and used for control and protection functions to stop the power system degradation. Successful implementation of real-time monitoring techniques based on synchrophasor technology demands the integration of several levels of triggering and

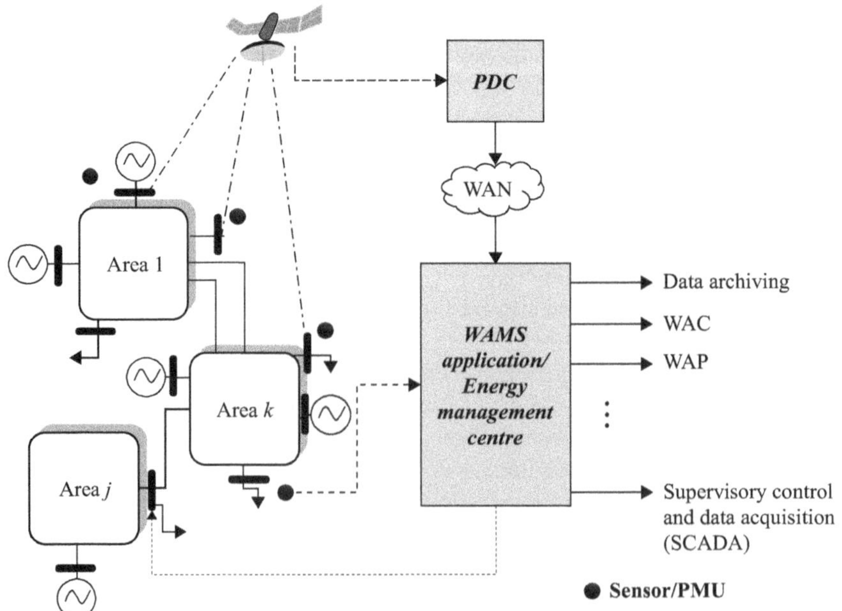

Figure 1.1 Generic WAMS

settings that detect deteriorating system conditions in the presence of normal power system behavior [9].

A diverse range of applications have been described for these technologies. Recent applications of these technologies include [3, 10–12]:

- Wide-area situational awareness
- State estimation
- Evaluation of security margins
- Real-time total transfer capability (TTC)
- Monitoring of inter-area oscillations
- Dynamic parameter identification
- Model validation
- Assessment of post-disturbance system integrity
- Phase angle monitoring
- Voltage stability monitoring
- Event-driven data archiving

The detection and characterization of temporal oscillations in measured data is greatly complicated by various factors. In practice, ambient and measurement noise, as well as impurities and artifacts may contaminate measured data, and lead to false alarms and erroneous operational decisions. Further, communication errors may cause data corruption and affect data analysis interpretation [13].

The design methods must incorporate both fault-tolerant strategies and data fusion techniques to enhance reliability and safety and also to improve the performance of wide-area monitoring systems.

Current WAMS architectures are evolving from advanced monitoring protection and control systems to more intelligent data fusion and control techniques. By taking advantage of advanced signal processing and data fusion methods, target tracking performance can be greatly increased.

1.3 Data collection and management

Significant developments in WAMS have originated from advanced data collection and processing followed by diagnosis and prognosis tools. Effective and reliable means of acquiring, managing, integrating, and interpreting performance data are needed for maximum useful information.

The term sensor can be interpreted broadly in this sense as a provider of high-quality synchronized data. This pre-processing stage can encompass various tasks such as data cleansing, dimensionality reduction, and removal of outliers.

Many technologies are used for observing system behavior. Many modern sensors are capable of simultaneously tracking various parameters. Examples include protection relays and disturbance fault recorders, frequency disturbance recorders in Internet-based frequency monitoring networks [14, 15]. Sensors, however, may provide an incomplete system observability, as PMU placement criteria are often application dependent [16].

While phasor measurement is often supplemented by SCADA measurements, this makes the measurement data heterogeneous and raises problems of synchronicity and alignment.

1.4 Challenges of future smart monitoring and analysis systems

Advanced power system measurement technologies have a key role to play in the development of Smart Grids [17, 18]. There are a number of issues that make the development of wide-area monitoring systems an extremely difficult and challenging task. Some of these issues are briefly discussed below.

Sensor selectivity and intelligent data fusion: Management of data is a key area of future research. Successful power system monitoring requires the efficient and reliable handling of large amounts of observational data. In general, data gathered by sensors is imperfect and may exhibit features that make the analysis of key system parameters difficult. Collected data from transient processes, in particular, is noisy and incomplete, heterogeneous, dynamically inconsistent, and may exhibit disparate time scales and spatio-temporal dependence.

The presence of outliers and noise, however, may lead to incorrect clustering and affect prediction performance since predictors attempt to predict the outliers and noise. Addressing these issues requires the development of specialized tools to systematically accommodate data collected across a wide range of spatial and temporal scales.

In addition, a complete (and reliable) view of the system cannot be achieved based on data from a single sensor. This highlights the need for the following:

Data paucity: Measurements are sparsely distributed over a large geographical area and may only provide partial observability of the system. Smart interpolation procedures are therefore needed to estimate system behavior at unsampled sites in space as well as to identify redundant measurement. Further, spatial measurements are spatially correlated (nonindependent over space) and techniques are needed to extract the true dimension of the measurement space.

The application of analytical methods to problems of system monitoring has been advanced greatly by the development of wide-area monitoring, protection, and control systems. The integration of large levels of renewable generation, however, presents significant challenges that must be addressed in designing the future smart grid.

Integrated communications across the grid: Emerging technological advances hold great promise for improving threat and health assessment [19, 20]. For example, new sensing techniques are being developed that scan system behavior at different rates and resolution. Integrating dynamically inconsistent data is a challenging task.

As more sensors are added to the power grid, the need arises to establish guidelines to place and manage system sensors in a coordinate manner. Such integration of process-based information with massive amount of data warrants further study. Further, determining the real-time data exchange needs is an open problem that has to be addressed.

Advanced sensing and metering: Advanced system monitoring is a key component of the future smart grid. Research on intelligent data analysis, machine learning and pattern techniques, and evolutionary computations are needed to make intelligent use of data [14–17]. Next-generation monitoring systems must effectively integrate and process data from various monitoring technologies while responding quickly and adaptively to changes in transient system behavior.

Sensor placement: As new (and more intelligent) sensors become available, efforts to improve placement of sensors are required. Since sensors are often placed based on various considerations, existing sensor networks may over-represent some regions and under-represent others. Moreover, as discussed earlier, site dispersion may not capture the spatial and temporal variability of the dynamic processes of interest.

Together these issues underline the need for a fast and robust method for adapting the following:

Analysis of incomplete data: Methods are needed for estimating system information from incomplete data sets. Filling in missing values with estimated values is a complex problem. Missing data may lead to unreliable or biased modal estimates and complicate the application of spatio-temporal models to system-wide data.

So far, no exhaustive analyses to study the nature of missing data have been reported. Modeling of heterogeneous data, however, promises to provide a better understanding of the problem since each type of data has its own advantages as well as limitations.

Bandwidth requirements: Other open problems include the following:

- Loss of PMU communication
- PMU streams with different reporting rates and different latency
- Fusing multimodal, multi-type data

References

1. John Hauer, Dan Trudnowski, Graham Rogers, Bill Mittelstadt, Wayne Litzenberger, Jeff Johnson, 'Keeping an eye on power system dynamics', *IEEE Computer Applications in Power*, 1997, pp. 50–54.
2. A. G. Phadke, 'The wide-world of wide-area measurement', *IEEE Power & Energy Magazine*, September/October 2008, pp. 52–65.
3. Mladen Kezunovic, Sakis Meliopoulos, Vaithianathan Venkatasubramanian, Vijay Vittal, *Applications of Time-Synchronized Measurements in Power Transmission Networks*, Power Electronics and Power Systems Series, Springer, Switzerland, New York, NY, 2014.

4. Jay Giri, Manu Parashar, Jos Trehern, Vahid Madani, 'The situation room – control center analytics for enhanced situational awareness', September/October 2012, pp. 24–39.
5. Damir Novosel, Vahid Madani, Bharat Bhargava, Khoi Vu, Jim Cole, 'Dawn of the grid synchronization – benefits, practical implementations, and deployment strategies for wide-area monitoring, protection, and control', *IEEE Power & Energy Magazine*, January/February 2008, pp. 49–60.
6. Miroslav Begovic, Damir Novosel, Daniel Karlsson, Charlie Henville, Gary Michel, 'Wide-area protection and emergency control', *Proceedings of the IEEE*, vol. 93, no. 5, May 2005, pp. 876–891.
7. S. Kincic, B. Wangen, W. A. Mittelstadt, M. Fenimore, M. Cassiadoro, V. VanZandt, L. Pérez, 'Impact of massive synchrophasor deployment on reliability coordination and reporting', 2012 IEEE Power and Energy Society General Meeting.
8. A. G. Phadke, J. S. Thorp, *Synchronized Phasor Measurements and their Applications*, Springer, New York, NY, 2008.
9. J. F. Hauer, F. Vakili, 'An oscillation detector used in the BPA power system disturbance monitor', *IEEE Transactions on Power Systems*, vol. 5, no. 1, February 1990, pp. 74–79.
10. Vladimir Terzija, Gustavo Valverde, Deyu Cai, Pawel Regulski, Vahid Madani, John Fitch, Srdan Sjok, ... Arun Phadke, 'Wide-area monitoring, protection, and control of future electric power networks', *Proceedings of the IEEE*, vol. 99, no. 1, August 2010, pp. 80–93.
11. Djordje Atanackovic, Jose H. Clapauch, Greg, Dwernychuk, Jim Gurney, 'First steps to wide area control – Implementation of synchronized phasors in control center real-time applications', *IEEE Power & Energy Magazine*, January/February 2008, pp. 61–68.
12. Jay Giri, David Sun, René Avila-Rosales, 'Wanted: A more intelligent grid', *IEEE Power & Energy Magazine*, March/April 2009, pp. 34–40.
13. Keith Holbert, Gerald T. Heydt, Hui Ni, 'Use of satellite technologies for power system measurements, command and control', *Proceedings of the IEEE*, vol. 93, no. 5, May 2005, pp. 947–955.
14. Z. Zhong, C. Xu, B. J. Billian, L. Zhang, S. J. S. Tsai, R. W. Conners, V. A. Centeno, ... Y. Liu, 'Power system frequency monitoring network (FNET) implementation', *IEEE Transactions on Power Systems*, vol. 20, no. 4, 2005, pp. 1914–1921.
15. Yingchen Zhang, Penn Markham, Tao Xia, Lang Chen, Yanzhu Ye, Zhongyu Wu, Zhiyong Yuan, ... Yilu Liu, 'Wide-area frequency monitoring network (FNET) architecture and applications', *IEEE Transactions on Smart Grid Applications*, vol. 1, no. 2, September 2010, pp. 159–167.
16. V. Madani, M. Parashar, J. Giri, S. Durbha, F. Rahmatian, D. Day, M. Adamiak, G. Sheble, 'PMU placement considerations – A roadmap for optimal PMU placement', 2011 IEEE/PES Power Systems Conference and Exposition (PSCE), Phoenix, 2011.

17. Pei Zhang, Fangxing Li, Navin Bhatt, 'Next generation monitoring, analysis, and control for the future smart control center', *IEEE Transactions on Smart Grid*, vol. 1, no. 2, September 2010, pp. 186–192.
18. Navin B. Bhatt, 'Role of synchrophasor technology in the development of a smarter transmission grid', 2010 IEEE Power and Energy Society General Meeting, Minneapolis, 2010.
19. North American Electric Reliability Corporation (NERC), 'Reliability considerations from the integration of Smart Grid', December 2010.
20. Stelios C. A. Thomopoulos, 'Sensor selectivity and intelligent data fusion', 1994 International Conference on Multisensor Fusion and Integration for Intelligent Systems, Las Vegas, 1994.

Chapter 2
Wide-area monitoring system architectures

2.1 Introduction

In recent years, a number of power utilities have designed and implemented advanced wide-area monitoring systems and data processing strategies to enhance grid stability and reliability. These strategies include the implementation of situational awareness systems including disturbance alert, event location triangulation and oscillation detection, early warning systems, data archiving, and other advanced features [1–3].

Wide-area monitoring systems will continue to evolve, as software, sensing and communications technology advance and signal processing tool improve. Various forms of wide-area monitoring systems have been developed to give early warning of power system disturbances [4, 5]. In order to integrate data from multiple sensors, specialized data fusion and communication techniques must be integrated into the existing wide-area monitoring systems (WAMS) architectures. The design methods must incorporate both fault-tolerant strategies and intelligent data fusion techniques to enhance reliability and safety and also to improve the performance of global monitoring systems.

This chapter gives a broad overview of various wide-area monitoring architectures, including sensor development and conceptualization, data processing and data fusion, and damage detection algorithms. Inspired by recent work on data fusion techniques, advanced WAMS architectures are introduced that represent a combination of measuring, monitoring, and analysis architectures. Key concepts in multisensor modeling, estimation, and fusion are introduced.

A framework for fusing data from multiple sensors to produce actionable intelligence is also identified. Advantages, challenging aspects, and recent advances in the design and implementation of WAMS architectures are reviewed. Areas where research is needed to advance the use of WAMS data are highlighted and described.

Challenges posed in developing distributed data fusion algorithms are also outlined.

2.2 WAMS architectures

Modern WAMS architectures provide a strategy to gather the data from different sensors and Phasor Data Concentrators (PDCs) and connect this information with

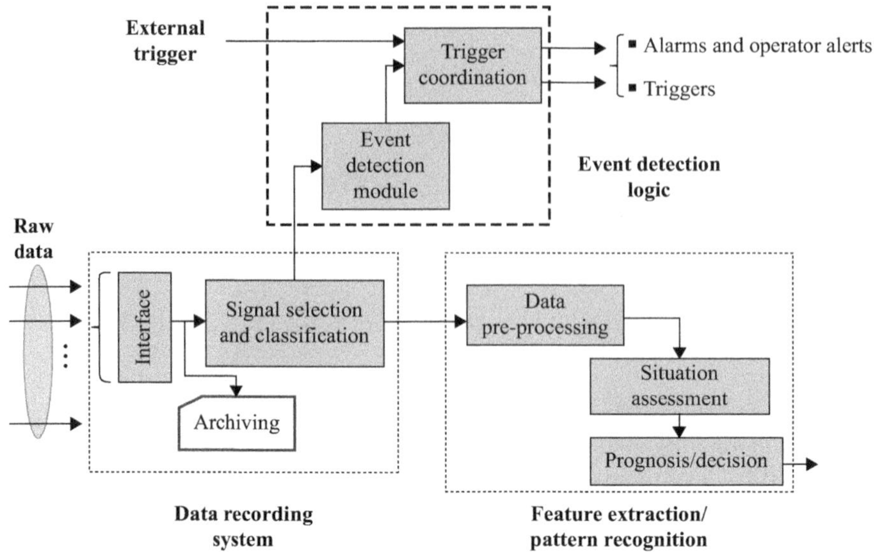

Figure 2.1 Generic WAMS structure

various data fusion algorithms and techniques to monitor power system oscillatory dynamics.

Various conceptualizations of the monitoring process exist in the literature. Figure 2.1 shows a generic three-layer WAMS structure inspired by [5]. The model includes the following three interrelated tasks:

1. Data acquisition and data management
2. Event logic detection and signal processing
3. Health monitoring and assessment

Discussion of the event logic detection and power health assessment modules is differed until Chapters 6 and 7. The issue of WAMS data management is discussed in sections 2.4 and 2.5.

The WAMS structure is typically hierarchical or distributed and can be split into two major levels [6, 7]: regional and global. At a local level, sensed information is automatically collected, synchronized, and archived by a monitoring and control center known as PDC. This information is then sent to a global data concentrator for real-time dynamic monitoring (RTDM), wide-area control (WAC), and wide-area protection (WAP). Inputs to the model may include raw data, past history, and expert knowledge [8].

Phasor measurement units (PMUs) collect and evaluate real-time data using advanced signal processing techniques. Recent advances in communication systems and digital electronics have enable the development of low-cost PMUs at distribution levels. As these technologies continue to advance, modeling and managing are becoming increasingly important.

In some applications the data may be subjected to pre-processing or pass directly into one of the other fusion levels. Pattern recognition techniques can then be used to identify features from the transient response as well as to discern the significant dynamic patterns containing dominant features in data. The ultimate goal is predictive modeling in which the behavior of unseen data is predicted for control and other purposes.

As pointed out in Chapter 1, PDCs can be used at a local, regional, or global level. Variations to these approaches are discussed in [6].

Monitoring provides critical data to be processed and used for control and protection functions to stop the power system degradation. Successful implementation of real-time monitoring techniques based on synchrophasor technology demands the integration of several levels of triggering and settings that detect deteriorating system conditions in the presence of normal power system behavior [9, 10].

Applications of these technologies include the following:

- State estimation
- Evaluation of security margins
- Real-time total transmission capability (TTC)
- Monitoring of inter-area oscillations
- Dynamic parameter identification
- Model validation
- Phase angle monitoring
- Voltage stability monitoring
- Event-driven data archiving

The detection and characterization of temporal oscillations in measured data is greatly complicated by various factors. In practice, ambient and measurement noise, as well as impurities and artifacts may contaminate measured data, and lead to false alarms and erroneous operational decisions [11].

WAMS architectures used in most power systems have gone through various developmental phases and gradually have evolved into two major types: centralized architectures and hierarchical or distributed architectures. The diversity of architectures and technologies that are used to collect and process measured data, however, is expanding rapidly.

2.2.1 Centralized WAMS architectures

Conventional WAMS architectures are centralized in nature. Figure 2.2 illustrates the centralized WAMS architecture, where measurements from all sensors or regional systems are processed globally in a central PDC. The structure consists of an ascending processing hierarchy in which local PDCs can be utilized to store and process information at low (global) levels. At the lower (raw) level, PDCs have basic applications in the order of a few PMUs. Archiving at a local level is necessary as transmitting large volumes of data as PMUs may take a substantial part of a substation bandwidth.

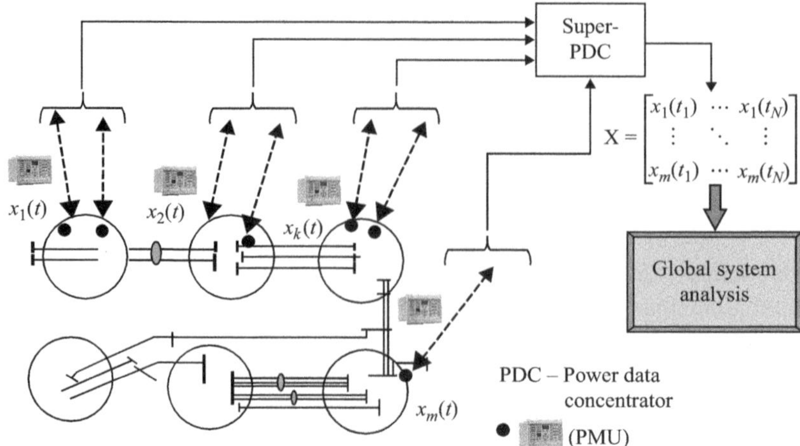

Figure 2.2 Centralized WAMS architecture

Advantages of centralized data integration include improved reliability and robustness, increase in both spatial and temporal coverage, and enhanced resolution. These approaches exploit shared structure but can result in a large amount of data to be monitored and concentrated at a single data concentrator, and the visualization of the multivariate data can be very difficult. Further, centralized architectures may be inflexible to sensor changes and the loss of packets of data may impact the quality of the received data and result in unreliable information.

With the increasing number of sensors in the system, data fusion algorithms can be quite complex. As suggested in Figure 2.2, data requirements coming from hundreds of PMUs may be very high. Networking constraints can be reduced by using especial communication strategies and advanced data fusion techniques.

Expert knowledge can also be used to determine relevant system behavior as well as to discriminate key system information.

By the nature of their construction, centralized WAMS structures can be both complex and expensive, and suit applications needing high levels of precision. The increasing complexity, sophistication, and size of modern power systems, coupled with a trend toward the use of distributed generation, argue more and more toward some form of distributed processing.

2.2.2 Hierarchical WAMS architectures

A conceptual view of this architecture is shown in Figure 2.3. In a decentralized WAMS architecture each PDC has its own local processor that can generally extract useful information from the raw sensor data prior to communication.

Decentralized structures are scalable and tolerant to dynamic changes in the network, but local information at a regional or local level has to be correlated with the information from other PDCs to extract global information. Further, separate analysis of each PDC data may not capture inter-area associations or result in a poor characterization of wide-area dynamics. Also, data fusion algorithms are complex.

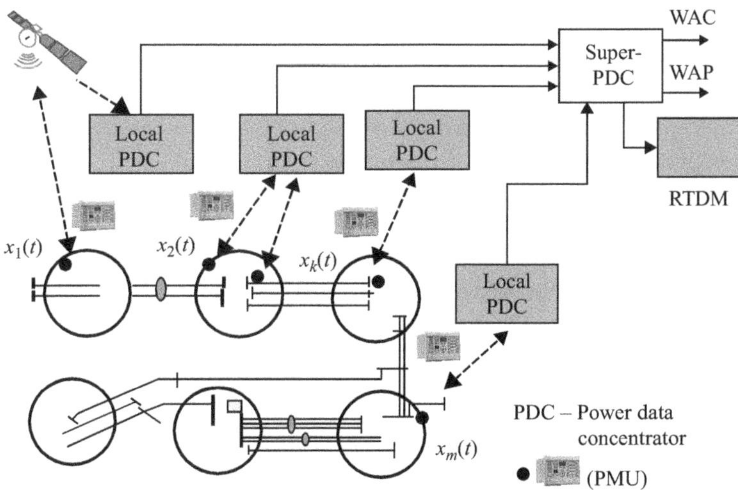

Figure 2.3 Decentralized control architecture

Examples of decentralized control structures that emulate modern data fusion architectures can be found in various recent applications [12–14].

2.2.3 Hybrid WAMS architectures

Recent advances in communication, computing, and sensing technologies have made it possible to develop hybrid WAMS architectures that take advantage of the relative strengths of centralized and decentralized architectures [15, 16].

In practice, WAMS architectures might include multiple hierarchical layers of PDCs that could perform local processing and archiving tasks.

2.3 Issues in data fusion

Modern WAMS incorporate multisensor data and information fusion techniques to enhance wide-area situational awareness and decision making. Data acquisition and data management are two key activities of real-time wide-area monitoring and analysis [16–18]. First, data must be combined and fused to obtain information of appropriate quality on which decisions can be made. This information can then be used for prognosis and decision.

This section introduces the problem of data analysis in the context of modern WAMS architectures and examines some sensing coverage aspects. Links to next-generation WAMS architectures based on smart sensors are discussed [17].

2.3.1 Data

At the core of a wide-area monitoring system is data. After the pre-processing stage, the data needs to be fused and analyzed at the feature or decision level.

The sections that follow introduce and specify the data fusion model and provide examples of how it operates.

2.3.2 Intelligent synchrophasor data fusion

Smart grid implementations require transmitting large sets of power system data. A typical data fusion architecture suitable for smart data fusion applications is shown in Figure 2.4, where data from multiple PDCs can be fused together for a more accurate characterization of system behavior.

Conceptually, data fusion encompasses a broad range of techniques from data pre-processing to feature extraction, data fusion, and decision support [19–20]. More advanced data fusion technologies also provide a framework for incorporating model errors and for updating model parameters or quantifying prediction uncertainties.

Key technologies for the smart data fusion include integrated communication across the grid, metering and measurement, and advanced data processing algorithms. The choice and implementation of the algorithms to process the data are one of the most crucial ingredients of a WAMS strategy.

As pointed out earlier, the monitored data can be processed and archived locally at each of the PDCs or globally at the super PDC to be compressed into lower dimensional data. In the latter case, the data may need to be compressed into lower dimensional data.

Dimensionality reduction at a local or global level is beneficial as it enables saving on the communication bandwidth and power required for transmitting data. Also, the approach reduces the burden on the global PDCs in the case of centralized data fusion architectures.

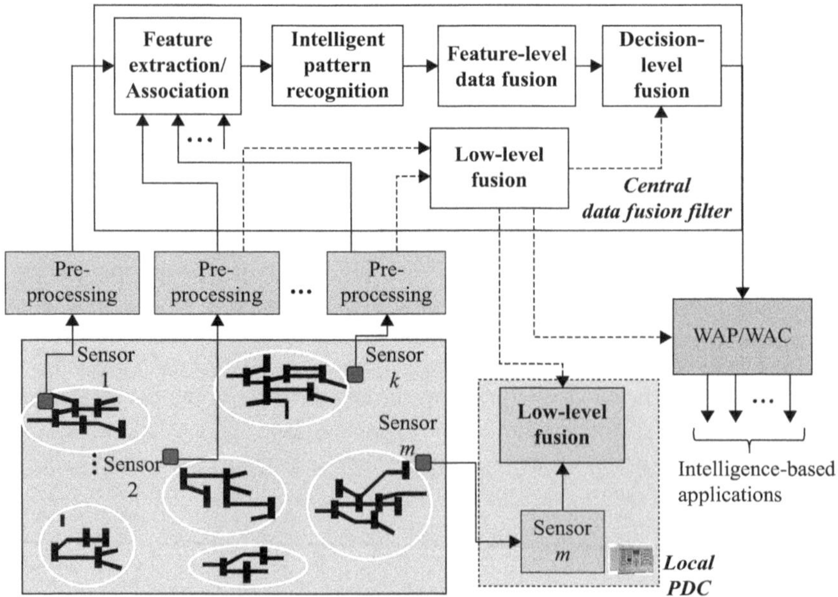

Figure 2.4 Hybrid data fusion strategy

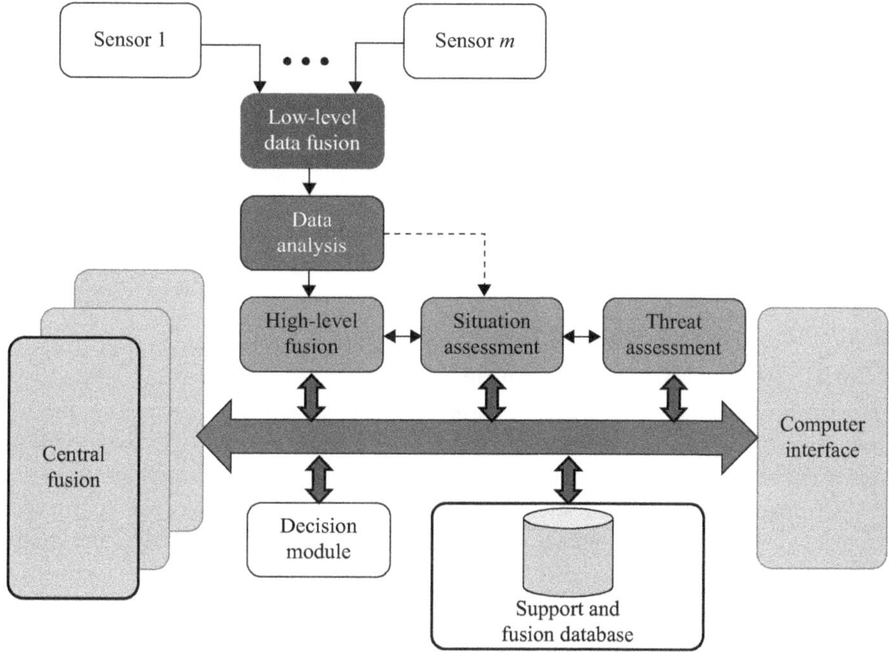

Figure 2.5 Data fusion architecture

Figure 2.5 provides a conceptual representation of the data fusion process. Key aspects of the data fusion/monitoring system include measurement, feature extraction, data association, and power system health assessment. These activities represent, in general, a closed-loop process with several levels of feedback.

There are different levels of data fusion. In general data fusion can be divided into three types: data oriented (low-level fusion), task oriented (feature extraction), and a mixture of data- and task-oriented fusion. A preprocessing stage is needed before fusing data from multiple sources:

Low-level fusion: Data fusion is performed before analysis. Examples of elementary fusion may include raw data averaging.

High-level fusion: Data fusion is performed after some data analysis. Typically, data reduction techniques are used to map data characteristics into distinct regions in the feature subspace through a process called feature extraction.

Feature-level fusion and data fusion: Data and features are fused together.

Data pre-processing: In general, before any data fusion can be performed the signals that come from PMUs should be pre-processed. Typical activities include normalization, filtering and amplification, error analysis, and noise treatment.

Modern sensors can monitor a wide variety of system parameters that can be used in various applications. In addition to power, voltage, frequency, phase angle, and tie-line power, PMUs can monitor other variables of interest such as rates of change of frequency and other parameters.

When combined with topological information, a spatio-temporal characterization of the system can be obtained that is no fully exploited in modern wide-area monitoring systems.

It is recognized that the potential of multiple data sets as well as their combination is not fully exploited. There are few analytical methods for manipulating and fusing information that have been developed. The following sections introduce key concepts in multisensor modeling and fusion.

2.3.3 Power system data fusion strategies

2.3.3.1 Formulation of the model

Data fusion is aimed to enhance situational awareness and decision making through the combination of information/data processing algorithms. A basic issue with the use of data fusion techniques is correlating multiple data sets.

Referring to Figures 2.3 and 2.4, consider a hierarchical WAMS architecture consisting of M areas or systems indexed $\{j = 1, \ldots, M\}$. To formalize the model, consider that each area has a network of m_k sensors, $\{k = 1, \ldots, M\}$ deployed to monitor system dynamic behavior and let the time evolution of a quantity of interest at sensor k be denoted by $x_k(t_j)$, $k = 1, \ldots, m_k$, $j = 1, \ldots, N$, where N is the number of observations. In addition, information about the state of switches is available.

To capture the space and time variability of transient processes, it is desirable to have an adequate distribution of the measurement sites as well as advanced signal processing techniques [21].

An illustration of this concept for a multiarea system is given in Figure 2.6. Three main cases can be considered:

1. Dense measurements – high-density sensor (PMU) networks at a local or regional (global) level
2. Sparse measurements – low-density (sparse) sensor networks
3. Lack of measurements

Each of these problem formulations is distinct and may require a different analysis approach. In the first case, measurements may be highly correlated and some sort of dimensionality reduction may be needed to avoid redundant information as well as to extract the true dimension of the data. Further, where measurements are inaccurate the estimated variables ought to be smooth containing no more structure than is warranted by the observation.

In the case of sparse sensor networks, spatio-temporal interpolation (prediction) may be used to estimate system behavior at unmonitored system locations. Extrapolation may also be needed to estimate values of physical variables in remote locations where they are not measured based on measurements taken at other system locations. These aspects are illustrated in Figure 2.6.

The advantage of this representation stems from its ability to model various WAMS architectures, including the centralized and decentralized architectures described above.

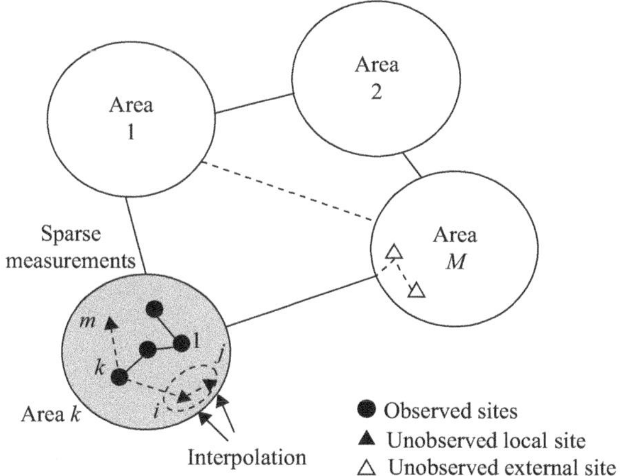

Figure 2.6 Multiarea power system showing dense sparse measurements at area k. Solid circles represent the location of measured observations. Estimates at unmonitored system locations can be obtained using a spatial interpolation method

In describing the adopted model, several concepts need to be made precise. The quantity or variable of interest (the measured data) will be denoted by x and may refer to different variables such as frequency and voltage measurements. In this context, this quantity refers to a data sequence. The observations can take on a variety of values, all of which are contained in the sample space.

Suppose that time histories of selected variables, $x_k(t_j)$, $k = 1, \ldots, m$, $j = 1, \ldots, N$, where N is the number of observations at m sensors, are simultaneously recorded in different temporal and spatial scales. Typically, the number of sampling times is much larger than the number of observing locations ($N \gg m$). A conceptual representation of the adopted model is shown in Figure 2.7.

In a centralized WAMS architecture, the observational data X is obtained directly from the PDCs with little or no processing.

In the distributed WAMS architecture, the set of local measurements can be arranged into an observation or data pattern matrix (refer to Figure 2.7)

$$\mathbf{X}_k = [\mathbf{x}_1 \quad \mathbf{x}_2 \quad \cdots \quad \mathbf{x}_{m_k}] = \begin{bmatrix} x_1(t_1) & \cdots & x_1(t_N) \\ \vdots & \ddots & \vdots \\ x_m(t_1) & \cdots & x_m(t_N) \end{bmatrix} \tag{2.1}$$

for $k = 1, \ldots, M$.

As discussed in subsequent chapters, each row of the observation matrix represents the time evolution of a given sensor. Each column can be thought of as a dynamic map.

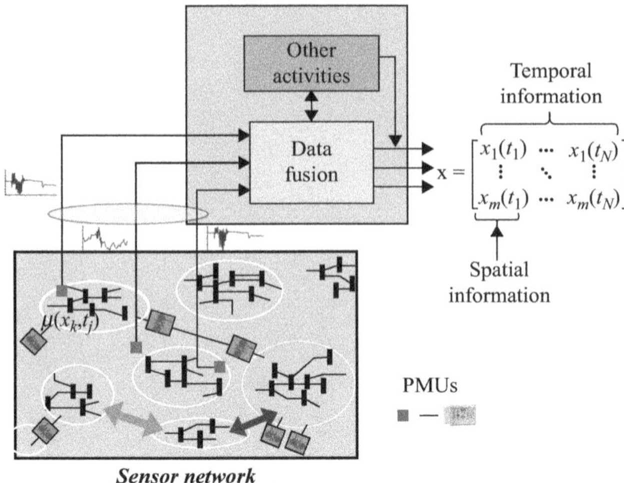

Figure 2.7 Definition of spatio-temporal information

Several interpretations of this model are possible:

1. Each row of the data matrix can correspond to a common variable, that is, frequency sensed at different buses.
2. Data can be heterogeneous involving power, frequency, phase angle, and voltage magnitudes.
3. The observational data matrix **X** can correspond to the same set of variables for different scenario events.
4. Data can include measured and historical or simulated data.

Other interpretations can be imagined and are introduced later in this book.

To pursue this concept further, assume that the data set obtained from each system is arranged into a global observation matrix, $\mathbf{X}_f$. Assembling the individual model for all areas, one can write

$$\mathbf{X}_f = [\mathbf{X}_1 \quad \mathbf{X}_2 \quad \cdots \quad \mathbf{X}_M] \tag{2.2}$$

where each of the block matrices is of the form (2.1), and it is assumed that all matrices have the same row dimension, as suggested in Figure 2.8.

Several issues arise with these representations. First, data can be highly correlated, especially in the case of dense measurements, and the true dimension is not known. Further, as the number of sensor increases, the problem of data handling becomes intractable. This leads to both, a dimensionality reduction problem and an intelligent fusion problem.

In addition, spatial coverage may be limited (sparse) or absent. In the first case, interpolation techniques are needed to estimate system behavior at unmeasured

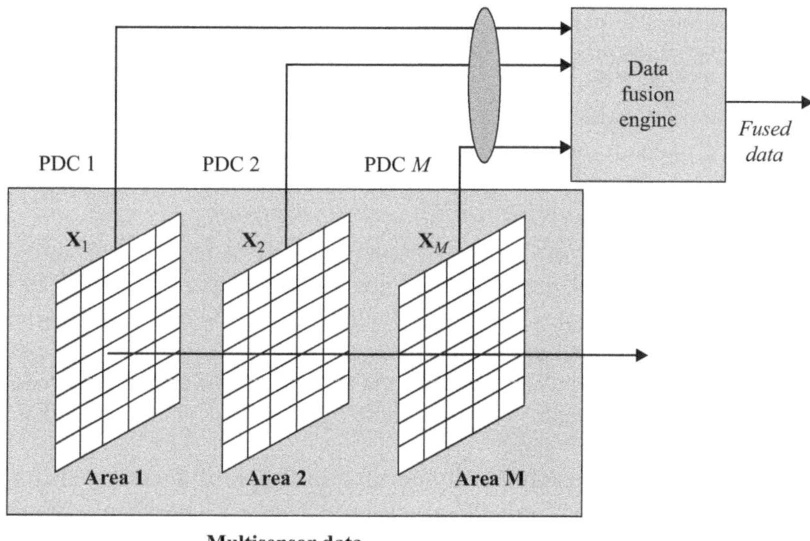

Figure 2.8 Elementary data fusion strategy

locations. In the latter case, measurements ought to be extrapolated from nearby measurement sites, as suggested in Figure 2.6.

Concatenating data in space or time may ignore the existence of mode of variations between individual observation matrices and may result in extremely large system models (2.2) that cannot be efficiently handled by current analysis techniques. Moreover, within-group variability is not reflected in the analysis.

Management of data and information for real-time performance monitoring of large power systems is a critical issue that warrants further investigation.

2.3.3.2 Hierarchical multiblock data models

Modern WAMS-generated data are typically extremely high dimensional and correlated in nature. For example, closely located sensors may generate highly correlated data. This means that the dimensionality of the data sets will be less than the number of sensors affecting the performance of some modal estimation algorithms.

An important problem in wide-area monitoring is the analysis and comparison of relationships between measures of system behavior or data sets. For example, in the analysis of inter-area oscillations, spatially and temporally measured data collected by the local networks of sensors at different geographical regions is used to estimate both local and wide-area dynamics under missing information, noise, or other effects. This analysis may also provide information about communication needs or local storage requirements. This problem has been largely neglected so far, with very few exceptions.

This problem can be cast as an intelligent synchrophasor data fusion problem and interpreted within the framework of modern approaches to optimize WAMS architectures [22, 23]. Reduction of variables often removes information and makes the interpretation misleading.

Chapter 5 expands on these ideas in the framework of a data fusion paradigm.

2.3.4 General framework for data assimilation

A related problem is that of feature (PDC-level) analysis involving the combined analysis of two data sets, for instance, associated with system regions or areas [19, 24]. Figure 2.9 illustrates schematically this idea. While it is straightforward to apply data analysis technique to each data set separately, the challenge lies in identifying common patterns across different data sets or relationships between or within data blocks. Examples might include the analysis of common inter-area modes or communication needs.

There are several ways in which one can use fusion or dimensionality reduction techniques to fuse data from different sensors or PDCs. One simple approach is to concatenate multiple data sets to aggregate common features, for example, through arithmetic averaging. Then, analysis can be performed in the common feature space to estimate common components. The results are then projected back into each data set to obtain individual components.

Techniques such as Principal Component Analysis (PCA), and Partial Least Squares (PLS), ensemble Kalman filtering, and intelligent neural methods can be used to monitor system behavior using synchronized phasor measurements [19, 24–27].

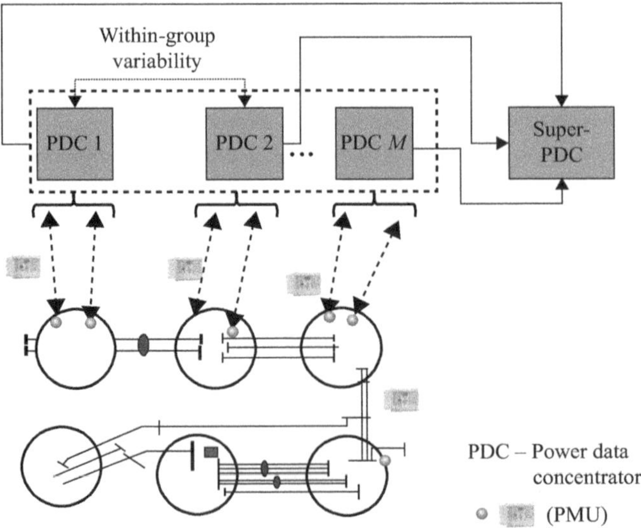

Figure 2.9 Partial least squares

In practice, in addition, the observation matrices are composed of a number of submatrices representing the internal groups within each area. Depending on the nature and dimensions of the individual observation matrices, the data are divided into blocks $\boldsymbol{X}_1, \boldsymbol{X}_2, \ldots, \boldsymbol{X}_M$:

$$\mathbf{X}_f = [\mathbf{X}_1 \quad \mathbf{X}_2 \quad \cdots \quad \mathbf{X}_M] \tag{2.3}$$

or

$$\mathbf{Y}_f = \begin{bmatrix} \mathbf{Y}_1 \\ \mathbf{Y}_2 \\ \vdots \\ \mathbf{Y}_M \end{bmatrix} \tag{2.4}$$

representing the local (area) dynamics. Each data matrix could correspond to nominal operating conditions or be a near real-time description of system behavior.

The data analysis problem is to relate the observational data matrix $\mathbf{Y}$ as some function of the data matrix $\mathbf{X}$. This allows to predict $\mathbf{Y}$ using the data of $\mathbf{X}$. A second objective is to extract relations between the data sets for system prognosis or detection.

In mathematical terms, the simplest approach is to assume a linear relation of the form

$$\mathbf{Y} = \mathbf{XC} + \mathbf{E} \tag{2.5}$$

where $\mathbf{C}$ is an unknown coefficient matrix.

One approach to the development of a flexible model that describes data of this form is adaptive multivariate statistical analysis such as partial least squares. Given two sets of observational data, $\mathbf{X}$ and $\mathbf{Y}$, the idea is to build an inner relationship or prediction of the form $\mathbf{U} = \mathbf{BT}$ where $\mathbf{B}$ is a matrix of coefficients and $\mathbf{T}$ is a matrix of scores or relationships. This entails the need to extend the applicability of data assimilation in modern WAMS architectures.

Based on where the assimilation process takes place, data fusion can be performed at different levels: local PDC level, regional PDC level, or super-PDC level. Each of these approaches requires a different treatment.

This approach is simple and easily adaptable to many different modeling problems. Other alternatives include neural nets and intelligent data fusion techniques.

A second approach would be to combine or fuse the data after applying a modal identification technique. This approach is illustrated schematically in Figure 2.10. Here, data is first decomposed into a series of modal components of the form

$$x_i^k(t) = \underbrace{\sum_{j=1}^{p_i} c_{ij}^k(t)}_{\substack{\text{Oscillatory} \\ \text{components}}} + \underbrace{m_i^k(t)}_{\text{Trend}} + \underbrace{\xi_i^k(t)}_{\text{Noise}}$$

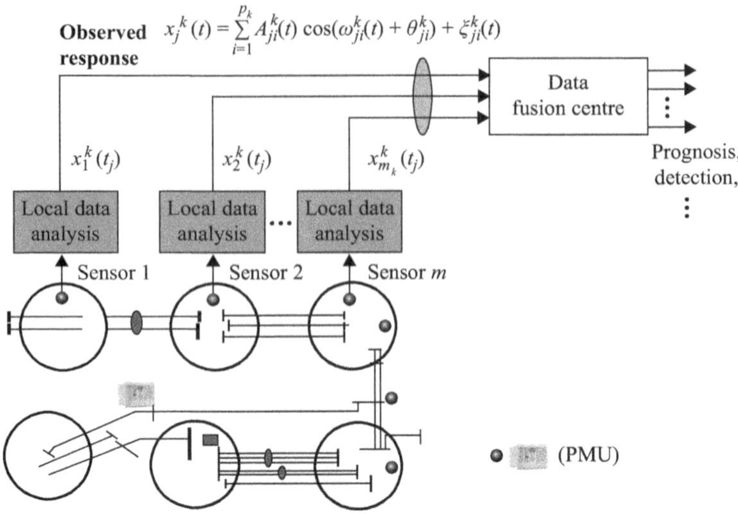

Figure 2.10 Conceptual view of feature-level fusion

where $c_{ij}^k(t)$ are oscillatory components, $m_i^k(t)$ is the time-varying mean, and $\xi_i^k(t)$ represents noise.

Examples of these formulations include modal decomposition methods such as Prony or wavelet analysis. The feature space can include modal parameters, trends, or relative quantities. A feature extraction technique will then be used to extract from the feature space–sensitive features.

Clearly, there are many other ways to perform data fusion and dimensionality reduction, which will be discussed in subsequent chapters.

2.4 Relationship between multiblock and single-block models

The earlier discussion leads to the notion of intelligent data fusion design which involves the design of advanced sensing and data processing strategies and architectures.

Analysis of multisensory data results, in general, in a three-way decomposition, which represents the time evolution of the data in terms of their spatial and time-dependent variations. A schematic illustration of this notion is shown in Figure 2.11 in which measured signals are first decomposed into modal components before data fusion.

Multivariate analysis techniques such as PCA, ICA, and multiblock partial least squares techniques have gained popularity in some scientific disciplines and are being developed and optimized for use in enhancing wide-area situational awareness and data management.

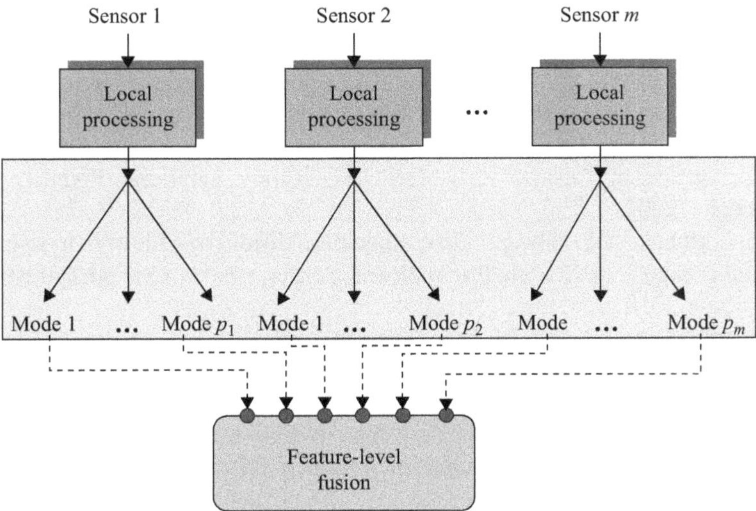

Figure 2.11 Conceptual view of feature-level fusion

This decomposition improves upon conventional spatio-temporal analyses in terms of greater accuracy, robustness, and computational speed.

These analysis techniques are explored further in subsequent chapters.

References

1. Daniel Karlsson, Morten Hemmingsson, Sture Lindahl, 'Wide-area system monitoring and control', *IEEE Power & Energy Magazine*, September/October 2004, pp. 68–76.
2. A. G. Phadke, J. S. Thorp, *Synchronized Phasor Measurements and their Applications*, Springer, New York, NY, 2008.
3. Vladimir Terzija, Gustavo Valverde, Deyu Cai, Pawel Regulski, Vahid Madani, John Fitch, Srdan Sjok, ... Arun Phadke, 'Wide-area monitoring, protection, and control of future electric power networks', *Proceedings of the IEEE*, vol. 99, no. 1, January 2011, pp. 80–93.
4. Damir Novosel, Vahid Madani, Bharat Bhargava, Khoi Vu, Jim Cole, 'Dawn of the grid synchronization – benefits, practical implementations, and deployment strategies for wide-area monitoring, protection, and control', *IEEE Power & Energy Magazine*, January/February 2008, pp. 49–60.
5. J. Hauer, D. J. Trudnowski, J. G. DeSteese, 'A perspective on WAMS analysis tools for tracking of oscillatory dynamics', 2007 IEEE Power Engineering Society General Meeting.
6. M. A. Weekes, B. A. Archer, 'Utility planning for a wide-area measurement system', IEEE PES General Meeting, 2009.

7. M. Kanabar, M. G. Adamiak, J. Rodrigues, 'Optimizing wide-area measurement system architectures with advances in phasor data concentrators (PDCs)', 2013 IEEE Power Engineering Society General Meeting.
8. Carlos Martinez, Henry Huang, Ross Guttromson, 'Archiving and management of power system data for real-time performance monitoring platform', Consortium of Electric Reliability Technology Solutions, PNNLL 15036, January 2005.
9. A. G. Phadke, J. S. Thorp, 'Communication needs for wide area measurement applications", IEEE 5th International Conference on Critical Infrastructure (CRIS), 2010.
10. Damir Novosel, Vahid Madani, Bharat Bhargava, Khoi Vu, Jim Cole, 'Dawn of the grid synchronization – benefits, practical implementations, and deployment strategies for wide-area monitoring, protection, and control', *IEEE Power & Energy Magazine*, January/February 2008, pp. 49–60.
11. Arturo Roman Messina, Vijay Vittal, Gerald Thomas Heydt, Timothy James Browne, 'Nonstationary approaches to trend identification and denoising of measured power system oscillations', *IEEE Transactions on Power Systems*, vol. 24, no. 4, November 2009, pp. 1798–1807.
12. Enrique Martinez, A. R. Messina, 'Modal analysis of measured inter-area oscillations in the Mexican interconnected system: The July 31, 2008 event', 2011 IEEE Power Engineering Society General Meeting.
13. Damir Novosel, Khoi Vu, Virgilio Centeno, Srdjan Skok, Miroslav Begovic, 'Benefits of synchronized measurement technology for power grid applications', *Proceedings of the 40th Hawaii International Conference on System Science*, Waikoloa, HI, 2007.
14. Mohammad Ilyas, Sami S. Alwakeel, Mohammed M. Alwakeel, *Sensor Networks for Sustainable Development*, CRC Press, Boca Raton, FL, 2014.
15. Yingchen Zhang, Penn Markham, Tao Xia, Lang Chen, Yanzhu Ye, Zhongyu Wu, Zhiyong Yuan, ... Yilu Liu, 'Wide-area frequency monitoring network (FNET) architecture and applications', *IEEE Transactions on Smart Grid Applications*, vol. 1, no. 2, September 2010, pp. 159–167.
16. Djordje Atanackovic, Jose H. Clapauch, Greg, Dwernychuk, Jim Gurney, 'First steps to wide area control – Implementation of synchronized phasors in control center real-time applications', *IEEE Power & Energy Magazine*, January/February 2008, pp. 61–68.
17. Pei Zhang, Fangxing Li, Navin Bhatt, 'Next generation monitoring, analysis, and control for the future smart control center', *IEEE Transactions on Smart Grid*, vol. 1, no. 2, September 2010, pp. 186–192.
18. US National Energy Technology Lab, 'Smart grid primer', 2006, http://energy.gov/oe/technology-development/smart-grid/smart-grid-primer-smart-grid-books.
19. Arturo R. Messina, Noé Reyes, Ismael Moreno, Marco A. Perez G., 'A statistical data-fusion-based framework for wide-area oscillation monitoring', *Electric Power Components and Systems*, vol. 42, nos. 3–4, 2014, pp. 396–407.

20. Junshan Zhang, Vijay Vittal, Peter Sauer, 'Networked information gathering and fusion of PMU data – Future grid initiative white paper', Power Systems Engineering Research Center, PSERC Publication 12-07, May 2012.
21. G. Giannuzi, D. Lauria, C. Pisani, D. Villaci, 'Real-time tracking of electromechanical oscillations in ENTSO-e continental European synchronous area', *International Journal of Electrical Power & Energy Systems*, vol. 64, January 2015, pp. 1147–1158.
22. M. Kanabar, M. G. Adamiak, J. Rodrigues, 'Optimizing wide area measurement system architectures with advancements in phasor data concentrators (PDCs)', 2013 IEEE Power and Energy Society General Meeting, Vancouver, BC, July 2013.
23. Anjan Bose, 'Smart transmission grid applications and their supporting infrastructure', *IEEE Transactions on Smart Grid*, vol. 1, no. 1, 2010, pp. 11–19.
24. Yuanjun Guo, Kang Li, D. M. Laverty, Loss-of-main monitoring and detection for distributed generations using dynamic principal component analysis', *Journal of Power and Energy Engineering*, vol. 2, 2014, pp. 423–431.
25. Bahador Khaleghi, Alaa Khamis, Fakhreddine O. Karray, Saiedeh N. Razavi, 'Multisensor data fusion: A review of the state-of-the-art', *Information Fusion*, vol. 14, 2013, pp. 28–44.
26. Carsten Montzka, Valentijn R. N. Pauwels, Harrie-Jan Hendricks Franssen, Xujun Han, Harry Vereecken, 'Multivariate and multiscale data assimilation in terrestrial systems: A review', *Sensors*, vol. 12, 2012, pp. 16291–16333.
27. Mladen Kezunovic, Sakis Meliopoulos, Vaithianathan Venkatasubramanian, Vijay Vittal, *Applications of Time-Synchronized Measurements in Power Transmission Networks*, Power Electronics and Power Systems Series, Springer, Cham, Switzerland, 2014.

Chapter 3
Spatio-temporal modeling of power system dynamic processes

3.1 Introduction

The development of wide-area monitoring systems provides unprecedented views of the system with increasing resolution and accuracy, coupled with capabilities of measuring new variables [1, 2]. Central to the development of advanced monitoring systems that improve the current predicting capabilities is the investigation of data correlations in both space and time.

Modeling of spatio-temporal measured data presents a unique set of problems as it often exhibits spatio-temporal dependence, nonlinearity, and heterogeneity [3–7]. To capture the space and time variability, it is desirable to have an adequate distribution of sensors. Different studies have incorporated spatial information within the framework of wide-area monitoring systems (WAMS) [3, 8–9]. Few attempts, however, have been made to integrate spatial and temporal approaches for investigating wide-area phenomena. In [8], spatio-temporal analysis methods have been used to examine dynamic trends and phase relationships between key system signals from measured data. These results, however, are intuitive; integrated spatio-temporal approaches are needed to quantify and understand patterns of system behavior in large-scale systems.

Predictions from spatio-temporal information from measured data can be used to study and evaluate patterns of variation in system dynamics, as well as to forecast the time evolution of transient processes [2]. Spatial models, in particular, may also serve as a basis for developing more complicated models that are better able to represent the observed oscillations at unmonitored locations [3]. Accurate spatial information can provide precise predictions of dynamic features such as propagating rates and the extent and distribution of mode propagation, and may improve the detection and tracking performance of monitoring systems [6].

In this chapter, promising new approaches for estimating and predicting a multi-variate spatio-temporal process from observational data are introduced. A general mathematical framework within which spatio-temporal analysis of large data sets can be performed is provided. In this approach, spatio-temporal models incorporating the dynamics of only a few temporal modes are developed. The technique can be used to monitor, model, and forecast wide-area physical processes such as inter-area oscillations and can reveal key features of a system.

Methods for extracting the dominant temporal and spatial components of variability in measured data are investigated and numerical issues are addressed.

3.2 Visualization of large space-time measurement data

Power system measured data are in the form of spatial time series, that is, time series of the same variable or sets of variables measured at a collection of locations [3]. Measured data can be conveniently interpreted in terms of spatio-temporal arrays.

Following Messina and Vittal [10], assume that $\mathbf{x}_k(t_j)$ denotes a sequence of observations of a measured transient process at locations x_k, $k = 1, \ldots, m$, and time t_j, $j = 1, \ldots, N$. The m locations represent sensors.

The data sets can be seen as a matrix in which each row represents the time evolution of a sensor, whose entries are the instantaneous measurements at time instance t. More formally, the time evolution of the transient process can then be described by the $m \times N$-dimension observation (snapshot) matrix $\mathbf{X}$:

$$\mathbf{X} = \begin{bmatrix} \mathbf{x}_1 & \mathbf{x}_2 & \cdots & \mathbf{x}_n \end{bmatrix}^T = \begin{bmatrix} x_1(t_1) & \cdots & x_1(t_N) \\ \vdots & \ddots & \vdots \\ x_m(t_1) & \cdots & x_m(t_N) \end{bmatrix} \tag{3.1}$$

where the superscript T indicates transpose, and typically $N > m$.

Physically, each row represents a time series, while each column can be seen as a map. A schematic depiction of the adopted system model is given in Figure 3.1. For purposes of analysis, variables are divided into observable (measured) and nonobservable (unmeasured).

The WAMS is developed as a data assimilation system and possesses the following capabilities:

1. Extracting from the observed response, the key dynamics of interest
2. Estimating system behavior at unmonitored system locations (unsampled points in space)

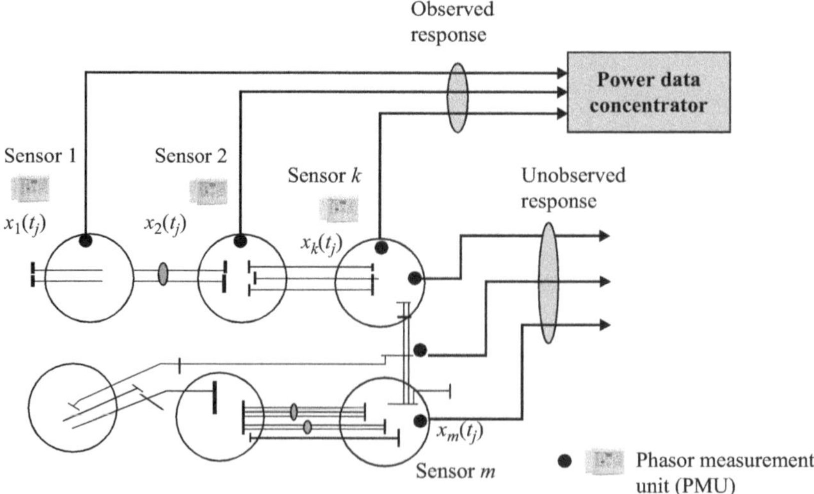

Figure 3.1 Measured system response

The first capability can be addressed using a suitable dimensionality reduction technique. The second capability requires the use of stochastic spatio-temporal models that can incorporate spatial relationships.

Dynamic information can be extracted on raw data or processed data using data fusion techniques as discussed in the foregoing chapters. Visualization methods can help to display the data, highlight their characteristics and reveal interesting characteristics

3.3 Spatio-temporal modeling of multivariate processes

Most practical process data contain contributions at multiple scales in time and frequency [9]. A key objective is to decompose such data into uncorrelated modes of variation and, specifically, in isolating these modes for further analysis, prognosis, and prediction [10–14].

The overall goal of multivariate methods for spatial time series is to represent the N-dimensional measured data by a decomposition of the form [15–17]

$$\mathbf{x}_k(t) = \sum_{j=1}^{p} a_j(t)\,\varphi_j(\mathbf{x}) + \varepsilon(t) \tag{3.2}$$

with $p < N$, where $a_1, a_2, \ldots, a_p$ is a set of temporal modes of variation, φ_j is a vector of dimension m that describes the spatial structure of the modes, and $\varepsilon(t)$ represents residual variability not captured by the p modes.

The approximation becomes exact for $p = \infty$. Particular cases of this model are wavelet analysis and proper orthogonal decomposition (principal component analysis, PCA). Discussion of more advanced techniques is postponed to later sections in this book.

Equation (3.2) is a prediction formula for $\mathbf{x}_k$ in which the first two terms on the *right hand side* (*rhs*) give the predicted value for $\mathbf{x}_k(t)$, and $\varepsilon(t)$ is the residual or prediction error. The basis function, φ, can be developed through a variety of techniques such as polynomials, trigonometric functions, or wavelets.

As an introduction to more advanced multivariate methods, the proper orthogonal decomposition (POD/PCA) method is introduced next.

3.3.1 *Empirical orthogonal function (EOF) analysis*

The method of POD is a model reduction technique that expands a set of data on empirically determined basis functions for modal decomposition. Given an array of measurements of the form (3.1), the technique aims at determining a vector φ that has the highest resemblance to all the observation vectors $\mathbf{x}$ simultaneously [18–20].

Let $\varphi_k(x_i)$, $k = 1, \ldots, l$, be an arbitrary set of orthonormal basis vectors (a low-order approximation), such that $\hat{\mathbf{x}}_i = \sum_{k=1}^{l} a_k(t)\,\varphi_k(\mathbf{x}_i)$.[1] The POD procedure

[1]For the finite-dimensional case $\mathbf{x}(t_j) = [x_1(t_j), x_2(t_j), \ldots, x_n(t_j)]^T$, the eigenfunctions $\varphi_k(\mathbf{x})$ are vector-valued functions of dimension n.

determines empirical orthogonal functions or proper orthogonal modes (EOFs or POMs), $\varphi_k(x)$, $k = 1, \ldots\infty$, such that the spatio-temporal data, $\mathbf{x}_k(t)$, at location k (a row of the observation matrix) and time t is approximated by the doubly orthogonal expansion [15]

$$\mathbf{x}_k(t) = a_o(t)\,\varphi_o(x) + \sum_{j=1}^{l} a_{kj}(t)\,\varphi_j(x) \tag{3.3}$$

in a least squares sense for each l, where the a_k are orthogonal, time-dependent amplitudes, the $\varphi_j(x)$ are spatial modes, and the term $a_o(t)$ captures the process mean.

More formally, POD analysis considers the problem of finding the orthonormal basis, φ_k, such that the Euclidean norm of the average error, ε_l,

$$\varepsilon(l) = \langle \| x_i(t_j) - \sum_{i=1}^{l} a_i(t)\,\varphi_i(x_i)\|^2 \rangle \tag{3.4}$$

is minimized for any $l < N$, where the notation $<.>$ denotes ensemble average.

This is equivalent to maximizing the quantity

$$\frac{1}{N(\varphi^T\varphi)}\left(\varphi^T\mathbf{X}\right)^2 = \frac{1}{N(\varphi^T\varphi)}\left(\varphi^T\mathbf{X}\right)\left(\varphi^T\mathbf{X}\right)$$

or

$$\frac{1}{N}\varphi^T\mathbf{X}\mathbf{X}^T\varphi = \frac{1}{N}\varphi^T\mathbf{R}\varphi \tag{3.5}$$

subject to the condition $\varphi^T\varphi = 1$, where $\mathbf{R} = \mathbf{X}\mathbf{X}^T$ is the m-by-m symmetric covariance or correlation matrix of the observations.

This leads to the eigenvalue problem

$$\mathbf{R}\varphi = \lambda\varphi \tag{3.6}$$

where $\lambda_1, \lambda_2, \ldots, \lambda_p$ are referred to as the POMs.

Using concepts from linear analysis, the error functional, $\varepsilon(l)$, can be expressed in the more useful form

$$\begin{aligned}\varepsilon(l) &= \sum_{i=1}^{N}\sum_{j=l+1}^{n} a_{ji}^2 = \sum_{i=1}^{N}\sum_{j=l+1}^{n} a_{ji}^2\left(\mathbf{x}_i^T\varphi_j, \mathbf{x}_i^T\varphi_j\right)\\ &= \sum_{j=l+1}^{n}\left(\mathbf{X}^T\varphi_j, \mathbf{X}^T\varphi_j\right) = \sum_{j=l+1}^{n}\varphi_j\mathbf{X}\mathbf{X}^T\varphi_j\end{aligned}$$

subject to $\varphi_i^T\varphi_j = \delta_{ij}$, $i, j = 1, \ldots, n$.

Properties of (3.6) have been studied by Messina and Vittal in the context of electromechanical modes [12].

The optimal POD basis vectors, $\varphi_j(x)$, can be found by introducing a smoothing parameter or Lagrange multiplier l_{ij} and solving the augmented system [10]

$$L = \sum_{j=l+1}^{n} \varphi_j^T \mathbf{X}\mathbf{X}^T \varphi_j - \sum_{i=l+1}^{n} \sum_{j=l+1}^{n} u_{ij}(\varphi_i^T \varphi_j - \delta_{ij}) \tag{3.7}$$

under the constraints $\varphi_i^T \varphi_j = \delta_{ij}$.

Physically, each spatial component map φ_j corresponds to a standing oscillation; the expansion coefficients a_{kj} represent how these patterns oscillate through time [11].

3.3.1.1 Physical interpretation

Writing (3.5) for all eigenvalues yields [18]

$$\boldsymbol{R}\Phi = \Lambda\Phi \tag{3.8}$$

where

$$\boldsymbol{\Phi} = [\varphi_1 \quad \varphi_2 \quad \cdots \quad \varphi_m]$$

$$\Lambda = \begin{bmatrix} \lambda_1 & & & \\ & \lambda_2 & & \\ & & \ddots & \\ & & & \lambda_m \end{bmatrix}$$

with $\boldsymbol{\Phi}^T\boldsymbol{\Phi} = \mathbf{I}$, where m denotes the number of measurement locations or sensors.

Combining (3.8) with (3.5), the relation between the measurement data and the modes can be expressed as

$$\frac{1}{N}\mathbf{X}\mathbf{X}^T\boldsymbol{\Phi} = \boldsymbol{\Phi}\boldsymbol{\Lambda} \tag{3.9}$$

or

$$\boldsymbol{\Phi}^{-1}\mathbf{X}\mathbf{X}^T\boldsymbol{\Phi} = N\boldsymbol{\Lambda} \tag{3.10}$$

where use has made of the property $\boldsymbol{\Phi}^T\boldsymbol{\Phi} = \mathbf{I}$.

Now defining $\mathbf{A}(t) = \boldsymbol{\Phi}^T\mathbf{X}$, one has that

$$\mathbf{A}\mathbf{A}^T = N\boldsymbol{\Lambda} \tag{3.11}$$

and

$$\mathbf{X} = \boldsymbol{\Phi}\mathbf{A} \tag{3.12}$$

where **A** is an m-by-N matrix of time-dependent coefficients of the form

$$\mathbf{A}(t) = \begin{bmatrix} \mathbf{a}_1(t) \\ \mathbf{a}_2(t) \\ \vdots \\ \mathbf{a}_m(t) \end{bmatrix} = \begin{bmatrix} a_{11}(t) & a_{12}(t) & \cdots & a_{1N}(t) \\ a_{21}(t) & a_{22}(t) & \cdots & a_{2N}(t) \\ \vdots & \vdots & \ddots & \vdots \\ a_{m1}(t) & a_{m2}(t) & \cdots & a_{mN}(t) \end{bmatrix}$$

with

$$\mathbf{a}_k(t) = [a_{k1}(t) \quad a_{k2}(t) \quad \cdots \quad a_{kN}(t)], \quad k = 1, \ldots, m$$

Using vector-matrix notation, this equation may be written more compactly as

$$\mathbf{X}(t) = \mathbf{\Phi}\mathbf{A}(t) = \mathbf{\Phi} \begin{bmatrix} \mathbf{a}_1(t) \\ \mathbf{a}_2(t) \\ \vdots \\ \mathbf{a}_m(t) \end{bmatrix} \tag{3.13}$$

The following points should be noted:

1. Each column of the data matrix (3.13) can be expressed in the form

$$\mathbf{x}_j = \sum_{i=1}^{m} a_{ij}\varphi_i, \quad j = 1, \ldots, N \tag{3.14}$$

2. POD analysis decouples spatial variability $\varphi(x)$, from temporal variability $a(t)$. More precisely:
   - Each spatial function $\varphi_j(x)$ is orthogonal to any other mode, that is

$$\varphi_k^T \varphi_j = \begin{cases} \delta_{kj} & \text{if } k = j \\ 0 & \text{if } k \neq j \end{cases}$$

   or

$$\mathbf{\Phi}^T\mathbf{\Phi} = \mathbf{I}$$

   - Each temporal function $a(t)$ is uncorrelated (orthogonal) to any other function:

$$\sum_{i=1}^{N} a_k(t_i)a_j(t_i) = \begin{cases} \lambda N & \text{if } k = j \\ 0 & \text{if } k \neq j \end{cases}$$

   or, equivalently

$$\mathbf{A}\mathbf{A}^T = N\mathbf{\Lambda}$$

Having computed an optimal low-dimensional representation of system dynamics, the distribution of each mode in each physical variable can be estimated using spatial prediction techniques.

3.3.1.2 Modal expansions

Once the statistical basis is determined using (3.3), the time evolution of the observed variables can be decomposed into a time-varying mean and a fluctuating part using (3.13) as

$$\begin{aligned}
\mathbf{x}_1 &= a_o(t)\,\varphi_o(x_1) + \sum_{k=1}^{p} a_{1k}(t)\,\varphi_k(\mathbf{x}) \\
\mathbf{x}_2 &= a_o(t)\,\varphi_o(x_2) + \sum_{k=1}^{p} a_{2k}(t)\,\varphi_k(\mathbf{x}) \\
&\;\;\vdots \\
\mathbf{x}_m &= a_o(t)\,\varphi_o(x_n) + \sum_{k=1}^{p} a_{mk}(t)\,\varphi_k(\mathbf{x})
\end{aligned} \tag{3.15}$$

These equations may be written more compactly as

$$\mathbf{x}_k(t) = \mathbf{x}_{av_k}(\mathbf{x}) + \mathbf{\Phi}\mathbf{a}_k(t) \tag{3.16}$$

where the first term on the *rhs* is the mean value, and the second term is the fluctuating part, with

$$\begin{aligned}
\mathbf{x}_{av}(\mathbf{x}) &= diag\left[a_o(t)\,\varphi_o(\mathbf{x}_1) \quad a_o(t)\,\varphi_o(\mathbf{x}_2) \quad \cdots \quad a_o(t)\,\varphi_o(\mathbf{x}_m)\right] \\
\mathbf{a}_k(t) &= \left[a_{1k}(t) \quad a_{2k}(t) \quad \cdots \quad a_{pk}(t)\right]^T
\end{aligned}$$

and

$$\mathbf{\Phi} = \lfloor \varphi_1(\mathbf{x}) \quad \varphi_2(\mathbf{x}) \quad \cdots \quad \varphi_p(\mathbf{x}) \rfloor$$

with $\mathbf{\Phi} = \lfloor \varphi_1(\mathbf{x}) \quad \varphi_2(\mathbf{x}) \quad \cdots \quad \varphi_p(\mathbf{x}) \rfloor$.

Motivated by this development, two relative measures are introduced

$$\hat{a}_i(t) = \frac{a_i(t)}{a_o(t)}, \quad i = 1, \ldots, p$$

and

$$\tilde{a}_i(t) = a_i(t) - a_o(t), \quad i = 1, \ldots, p$$

Extensions and generalizations to these basics are discussed later in this book. Appendix A provides a physical interpretation of proper orthogonal modes and provides links to other formulations.

3.3.2 SVD-based proper orthogonal decomposition

A useful alternative to POD analysis can be obtained from singular value decomposition (SVD) analysis of the data matrix. In this framework, the SVD of (3.1) is given by [21, 22]

$$\mathbf{X} = \mathbf{U\Sigma V}^T = \mathbf{U}[\,\mathbf{\Sigma}_m \quad \mathbf{0}\,]\begin{bmatrix}\mathbf{V}_1^T \\ \mathbf{V}_2^T\end{bmatrix} \tag{3.17}$$

where $\mathbf{U}$ is the m-by-m orthonormal matrix containing the left singular vectors, $\mathbf{V}$ is an N-by-N matrix containing the right singular vectors, Σ is an m-by-N matrix containing the singular values σ, defined as

$$\mathbf{\Sigma} = \begin{bmatrix} \sigma_1 & & & \\ & \sigma_2 & & \\ & & \ddots & \\ & & & \sigma_p \end{bmatrix}$$

Comparison of (3.17) and (3.13) shows that

$$\begin{aligned} \mathbf{A} &= \mathbf{\Sigma V}^T \\ \mathbf{\Phi} &= \mathbf{U} \end{aligned} \tag{3.18}$$

and

$$\begin{aligned} \mathbf{A} &= \mathbf{U}^{-1}\mathbf{X} = \mathbf{U}^T\mathbf{X} \\ \mathbf{X} &= \mathbf{UA} \end{aligned}$$

as expected.

A physical interpretation of the method is provided in [23].

3.3.2.1 Prediction

The above framework may be extended to modeling large data sets. Let the singular value decomposition of the data matrix, $\mathbf{X}$, be written as

$$\mathbf{X} = \mathbf{U\Sigma V}^T = \mathbf{U}\Lambda^{1/2}\mathbf{V}^T \tag{3.19}$$

where $\Sigma = \text{diag}[\sigma_1\ \sigma_2\ \ldots\ \sigma_r]$is a diagonal matrix whose entries are the singular values of $\mathbf{X}$, and the columns of $\mathbf{U}$, $\mathbf{V}$ are the left singular vectors and right singular vectors of $\mathbf{X}$, respectively.

The square matrix $\Lambda^{1/2}$ has entries defined by

$$\mathbf{\Lambda} = \begin{bmatrix} \sqrt{\lambda_1} & & & \\ & \sqrt{\lambda_2} & & \\ & & \ddots & \\ & & & \sqrt{\lambda_n} \end{bmatrix}$$

A straightforward analysis reveals that [21]

$$\begin{cases} \mathbf{XX}^T = \mathbf{U\Sigma}^2\mathbf{U}^T \\ \mathbf{X}^T\mathbf{X} = \mathbf{V\Sigma}^2\mathbf{V}^T \end{cases}$$

As a consequence, the left and singular vectors of $\mathbf{X}$ are the eigenvectors of $\mathbf{XX}^T$ and $\mathbf{X}^T\mathbf{X}$, respectively, and $\sigma_i = \sqrt{\lambda_i}$.

Let now the eigenvalues and eigenvectors of the covariance matrix $\mathbf{XX}^T$ be $\lambda_1 \geq \lambda_2 \geq \lambda_r \geq \lambda_{r+1} = \ldots = \ldots \lambda_N = 0$ and $\mathbf{V} = [\mathbf{V}_1\ \ \mathbf{V}_2]$, respectively, where $\mathbf{V}_1 = [\mathbf{v}_1, \mathbf{v}_2, \ldots, \mathbf{v}_r]$ and $\mathbf{V}_2 = [\mathbf{v}_{r+1}, \mathbf{v}_{r+2}, \ldots, \mathbf{v}_N]$, where r is the index of the smallest positive eigenvalue of $\mathbf{X}\mathbf{X}^T$.

From (3.19), it can easily be verified that

$$\mathbf{X}^T\mathbf{X}\mathbf{V} = \mathbf{V}\boldsymbol{\Sigma}^2$$

and

$$[\mathbf{V}_1 \,\vdots\, \mathbf{V}_2]^T\mathbf{X}\mathbf{X}^T[\mathbf{V}_1 \,\vdots\, \mathbf{V}_2] = \left[\begin{array}{c:c} \boldsymbol{\Sigma}_1^2 & 0 \\ \hdashline 0 & \boldsymbol{\Sigma}_2^2 \end{array}\right]$$

where it is assumed that the σ_i are the singular values of $\mathbf{X}^T$.

Defining now $\mathbf{U} = [\mathbf{U}_1\ \mathbf{U}_2]$, where $\mathbf{U}_1 = [\mathbf{u}_1, \mathbf{u}_2, \ldots, \mathbf{u}_r]$ and $\mathbf{U}_2 = [\mathbf{u}_{r+1}, \mathbf{u}_{r+2}, \ldots, \mathbf{u}_N]$, yields

$$\mathbf{U}^T\mathbf{X}^T\mathbf{V} = \begin{bmatrix} \mathbf{U}_1^T\mathbf{X}^T\mathbf{V}_1 & \mathbf{U}_1^T\mathbf{X}^T\mathbf{V}_2 \\ \mathbf{U}_2^T\mathbf{X}^T\mathbf{V}_1 & \mathbf{U}_1^T\mathbf{X}^T\mathbf{V}_2 \end{bmatrix} = \begin{bmatrix} \boldsymbol{\Sigma}_1 & \mathbf{0} \\ \mathbf{0} & \boldsymbol{\Sigma}_2 \end{bmatrix}$$

After some manipulations, it can be shown that

$$\mathbf{X} = \overbrace{\boldsymbol{\Lambda}_{av}}^{\text{Temporal mean}} + \mathbf{U}_1 \overbrace{\boldsymbol{\Sigma}_1\mathbf{V}_1^T}^{\text{Principal components}} \tag{3.20}$$

and

$$\mathbf{X} = \overbrace{\boldsymbol{\Gamma}_{av}}^{\text{Spatial mean}} + \mathbf{U}_2 \overbrace{\boldsymbol{\Sigma}_2\mathbf{V}_2^T}^{\text{Principal components}} \tag{3.21}$$

where Γ_{av} and Λ_{av} are matrices containing the temporal and spatial means, and the products $\Sigma\mathbf{V}^{\mathrm{T}}$ are called the principal components.

Of note, the second term in (3.20) can be further decomposed in the form

$$\mathbf{U}_1\boldsymbol{\Sigma}_1\mathbf{V}_1^T = \overbrace{\mathbf{T}\mathbf{P}^T}^{\text{Important variation}} + \overbrace{\mathbf{E}}^{\text{Unimportant variation}}$$

where the first term on the *rhs* contains the important components making up the observations, and the second term on the *rhs* is a residual matrix.

By detrending the observation matrix $\mathbf{X}$, a first measure of power system oscillatory behavior is obtained from the time-demeaned matrix

$$\mathbf{X}_{osc} = \mathbf{X} - \boldsymbol{\Lambda}_{av} \tag{3.22}$$

Large deviations from the global mean are a key indicator of system deterioration. In addition, the mode shapes associated with critical modes provide information related to the extent and distribution of system damage.

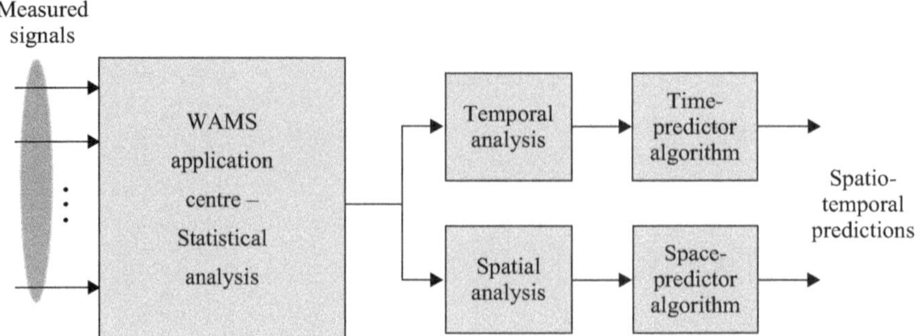

Figure 3.2 Spatio-temporal prediction framework

Figure 3.2 illustrates the notion of spatial and temporal prediction. The goal of employing spatio-temporal models is to explicitly account for the effects of site dispersity in modal estimates.

3.3.2.2 Entropy and energy

From (3.13) two basic types of characteristic quantities can be derived: entropy and energy. Using basic energy principles, the total fluctuating energy can be expressed at a given time instant t as [24]

$$E(t) = \sum_{k=1}^{p} \mathbf{a}_k^T(t)\mathbf{a}_k(t) = \sum_{k=1}^{p} E_k(t) \tag{3.23}$$

where $E_k(t)$ is the energy captured by the kth mode, and second and higher-order contributions are neglected.

The portion of energy in each mode is thus defined by [25]:

$$p_k(t) = E_k(t) / \sum_{k=1}^{p} E_k(t)$$

where the term p_k is a probability. As discuss below, this allows to define entropy in terms of energy.

3.3.3 *Departure from mean value*

In practical applications, the measured time histories are written as the sum of their mean and fluctuating components as follows. Let the mean value for each time series be stacked in the vector

$$\mathbf{x}_{mean} = \begin{bmatrix} \mathbf{x}_{mean_1} & \mathbf{x}_{mean_2} & \cdots & \mathbf{x}_{mean_m} \end{bmatrix}^T$$

where

$$x_{mean_j}(t) = \frac{1}{N}\sum_{k=1}^{N} x_k(t)$$

It then follows that the deviations from the mean value are given by

$$\mathbf{X} = \mathbf{X} - \mathbf{1}_m \mathbf{x}_{mean} = \mathbf{H}\mathbf{X} \tag{3.24}$$

where $\mathbf{1}_m$ is a vector of dimension m with all elements unity, and the superscript T denotes transpose, and $\mathbf{H}$ is a centering matrix of order m, defined as

$$\mathbf{H} = \left(\mathbf{I}_m - \frac{1}{m}\mathbf{1}_m\mathbf{1}_m^T\right)$$

Motivated by this development, a procedure to estimate modal information at unsampled system locations is now suggested.

3.4 Spatio-temporal interpolation methods

Spatial interpolation is defined as the prediction of the unknown value of a physical variable from known measurements obtained at a set of sample locations [25]. These methods have been briefly reviewed before in the context of the present analysis.

3.4.1 Background

Following the notation of the previous section, let $[x_1, x_2, \ldots, x_m]$ represent a set of observed (measured) system data at locations $\{1, 2, \ldots, m\}$, and let x_i be the unknown data at an unmonitored location x_i within a study area, as shown in Figure 3.3.

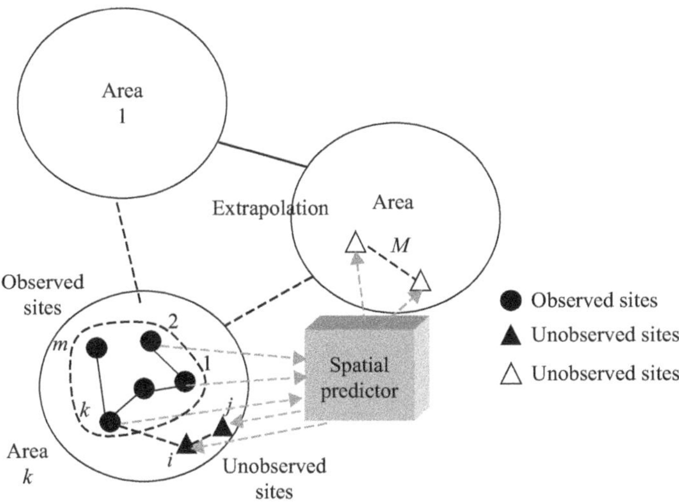

Figure 3.3 Illustration of the problem of monitoring. Solid circles indicate the location of measurement points. Solid triangles and empty circles indicate unmonitored system locations inside and outside of the study area, respectively

Formally, the problem of spatial interpolation can be defined as follows. Given a set of observed data, an estimate value of the random process at an unobserved site x_i at time t_j can be obtained from [26]

$$\hat{x}_i(t) = f\left(x_1(t), x_2(t), \ldots, x_p(t)\right) \tag{3.25}$$

where f represents a suitable spatial interpolation functional, and p represents the number of sampled points used for the estimation.

The functional form of f depends on the particular interpolation method. An estimate value (a prediction) $\hat{x}_i(t)$ of the random process at an unobserved site x_i at time t_j can be obtained from a weighted average of values at sample points as [27, 28]

$$\hat{x}_i(t) = \sum_{l=1}^{m} w_{il} x_l(t_j) = \sum_{l=1}^{m} w_{il} \sum_{k=1}^{p} a_k(t)\, \varphi_k(\mathbf{x}) \tag{3.26}$$

in which i is the location of the estimate, l is a sensor location, and the terms w_{il} are the unknown prediction weights (weighting functions) for the measurement site l and the unobserved site i; m is the number of nearby stations (sensors) that influence the estimate at location i.

Using the notion of similarity, the weight w_{il} for the unmonitored site l and the sampled site i can be defined as [29]

$$w(x_i, x_l) = w_{il} = \frac{s(x_i, x_l)}{\sum_{l=1}^{m} s(x_i, x_l)} \tag{3.27}$$

where $s(x_i, x_l)$ is a similarity coefficient used to quantify the degree of resemblance or affinity between sites i and l with respect to a set of auxiliary (measured) variables.

Several variations to this representation are possible and are discussed below. Figure 3.3 illustrates the proposed model. Two estimation problems are considered:

1. Prediction of system dynamics at an unmonitored site within the study area (spatial interpolation)
2. Prediction at unmonitored locations outside the study region (spatial extrapolation)

The treatment of the second issue requires the development of techniques to generate spatial structures.

3.4.2 Similarity measures

A flexible approach that provides a framework to capture spatial structures and their time evolution is spatial interpolation. Consider again a set of measurement sites $\{x_1, x_2, \ldots, x_m\}$.

Following de Jong *et al.* [30] these measurement points and their relations form a network described by a spatial weighting matrix **C** that indicates the existence of a relation between points i and j.

More formally, a distance matrix, **C**, can be defined as a matrix containing zeros, except at the interconnection of neighboring observations, which contain pairwise Euclidean distance coefficients, c_{ij}, that is

$$\mathbf{C} = [C_{ij}] = \begin{cases} c_{ij} = |\min P_{ij}| & \text{if } i \neq j \\ 0 & \text{if } i = j \end{cases} \tag{3.28}$$

where c_{ij} is the distance between locations i and j, and P_{ij} is the shortest path between vertexes i and j. Typically, values are assumed to be normalized such that c_{ij} ranges from 0 to 1.

Different functions $f(c_{ij})$ are available including exponential, spherical and Gaussian as discussed below. Particular cases are binary connectivity matrices [31], and distance matrices used in voltage stability studies [32].[2]

Following Borcard and Legendre [31], a similarity matrix **S** can be defined as

$$\mathbf{S} = [s_{ij}] = \begin{bmatrix} s_{11} & s_{12} & \cdots & s_{1n} \\ s_{21} & s_{22} & \cdots & s_{2n} \\ \vdots & \vdots & \ddots & \vdots \\ s_{n1} & s_{n2} & \cdots & s_{nn} \end{bmatrix} \tag{3.29}$$

with

$$s_{ij} = 1 - \left(\frac{c_{ij}}{\max(c_{ij})}\right)^2$$

where the similarity coefficients, s_{ij} range from 0 (for $d_{ij} = max(d_{ij})$) to 1 (for $d_{ij} = 0$) and provide a measure of the strength of the connection.

Alternatively, (3.29) can be rewritten in the more useful form

$$\mathbf{S} = [s_{ij}] = \mathbf{1}_n \mathbf{1}_n^T - \frac{\mathbf{C}^*}{\max(c_{ij}^*)^2} \tag{3.30}$$

where $c_{ij}^* = (d_{ij})^2$, $s_{ij} = 1 - (d_{ij})^2/\max(d_{ij})^2$, $\mathbf{1}_n$ is a vector of dimension n with all elements unity, and the superscript T denotes transpose. Similarity matrices can be interpreted as weighted graphs where the intensity of the connections is given by the coefficients s_{ij}.

Once, the similarity coefficients are computed, spatial weights and bases functions can be obtained using (3.15).

[2]A binary connectivity matrix, **C**, is defined as a matrix containing zeros, except at the interconnection of neighboring observations, which contain ones.

3.4.3 Spatial structures

A convenient measure to compare the value of a measurement at any one location $u(x_i,t)$ with the values at all other locations (i.e., to test for global spatial autocorrelation) is the global Moran's autocorrelation coefficient [17]. Consider an m-dimensional vector $\mathbf{x} = [x_1, \ldots, x_m]^T$, containing measurements of a variable of interest at n sites, and let $\mathbf{C}$ be a mxm symmetrical spatial weighting (distance) matrix. The Moran coefficient $I(\mathbf{x})$ is defined as [18]

$$I(\mathbf{x}) = \frac{m}{2(0.5\sum_{i=1}^{m}\sum_{j=1}^{m} C_{ij})} \frac{\sum_{i=1}^{m}\sum_{j=1}^{m} z_i C_{ij} z_j}{\sum_{i=1}^{n} z_i^2} \\ = \frac{m}{\mathbf{1}_m^T \mathbf{C} \mathbf{1}_m} \frac{\mathbf{x}^T(\mathbf{I}_m - \mathbf{1}_m\mathbf{1}_m^T/m)\,\mathbf{C}(\mathbf{I}_m - \mathbf{1}_m\mathbf{1}_m^T/m)\mathbf{x}}{\mathbf{x}^T(\mathbf{I}_m - \mathbf{1}_m\mathbf{1}_m^T/m)\mathbf{x}} \tag{3.31}$$

where m is the number of observation locations, $z_i = x_i - \sum_{j=1}^{m} x_j/m$ is the sample mean, and $\mathbf{I}_m$ is the mxm identity matrix. The index ranges from $+1$ to -1.

Physically, the spatial correlation coefficient (3.13) indicates to what extent the observations x_i, $i = 1, \ldots, m$, influence each other via the structure of the network. As discussed in [16], the Moran coefficient is positive when the observed measurements of locations within the distance tend to be similar, negative when they tend to be dissimilar, and approximately zero when they are arranged randomly and independently over space [16]. High values of the coefficient indicate that autocorrelation is high.

It can be easily verified that the eigenvectors of the spatial weight matrix $\mathbf{C}_c = (\mathbf{I}_m - \mathbf{1}_m\mathbf{1}_m^T/m)\,\mathbf{C}(\mathbf{I}_m - \mathbf{1}_m\mathbf{1}_m^T/m)$ are mutually orthogonal and uncorrelated. It can be shown [18] that the upper and lower values of the autocorrelation coefficient $I(\mathbf{x})$ are given by $(\mathbf{I}_m - \mathbf{1}_m\mathbf{1}_m^T/m)\lambda_{\max}$ and $(\mathbf{I}_m - \mathbf{1}_m\mathbf{1}_m^T/m)\lambda_{\min}$, where $\lambda_{\max}$ and $\lambda_{\min}$ are the extreme values of matrix $\mathbf{C}_c$.

Also of relevance, the eigenvectors of matrix $\mathbf{C}_c$ form orthogonal sets of spatial structure; eigenvectors associated with large positive eigenvalues describe global structures while those associated with negative values give an indication of local structures.

Despite its simplicity, the connectivity approach has rich mathematical structure and may provide a fruitful way of representing and studying a variety of issues in network.

3.4.4 Derivation of weights

Let the estimates of key system variables at unsampled locations be given by (3.11). Based on (3.13), a spatial map of eigenvectors can be obtained from the eigendecomposition of the spatial weighting matrix $\mathbf{C}_c$. Then, an estimate of the temporal coefficients (weights) is obtained from EOF analysis.

An outline of the algorithm is as follows:

Computation of spatial weights

1. Compute the distance matrix **C** using (3.28). Compute the singular values s_i, and singular vectors **U** from $\mathbf{C}_c$.
2. Using the most significant modes of the spatial structure, compute the weights w_{ij} for (3.27) as

$$w_{ij} = \sum_{k=1}^{p} \varphi_{ik}(\mathbf{x})^T \mathbf{a}_{ik}(t) \tag{3.32}$$

3. Once the weights are computed, estimates of system behavior at unsampled sites are computed from (3.26) as

$$\begin{aligned}
\hat{x}_1(t) &= \sum_{k=1}^{p} \varphi_k(\mathbf{x}_1)^T \mathbf{a}_{1k}(t) \left[\sum_{i=1}^{N} w_{i1} \right] \\
\hat{x}_2(t) &= \sum_{k=1}^{p} \varphi_k(\mathbf{x}_2)^T \mathbf{a}_{1k}(t) \left[\sum_{i=1}^{N} w_{i2} \right] \\
&\vdots \\
\hat{x}_r(t) &= \sum_{k=1}^{p} \varphi_k(\mathbf{x}_n)^T \mathbf{a}_{1k}(t) \left[\sum_{i=1}^{N} w_{in} \right]
\end{aligned} \tag{3.33}$$

or

$$\begin{bmatrix} \hat{x}_1(t) \\ \hat{x}_2(t) \\ \vdots \\ \hat{x}_m(t) \end{bmatrix} = \begin{bmatrix} w_{11} & w_{12} & \cdots & w_{1p} \\ w_{21} & w_{22} & \cdots & w_{2p} \\ \vdots & \vdots & \ddots & \vdots \\ w_{m1} & w_{m2} & \cdots & w_{mp} \end{bmatrix} \begin{bmatrix} \sum_{k=1}^{p} \varphi_k(\mathbf{x})^T \mathbf{a}_{1k}(t) \\ \sum_{k=1}^{p} \varphi_k(\mathbf{x})^T \mathbf{a}_{2k}(t) \\ \vdots \\ \sum_{k=1}^{p} \varphi_k(\mathbf{x})^T \mathbf{a}_{pk}(t) \end{bmatrix}$$

The algorithm is simple to implement, and computational requirements are small.

3.4.5 *Practical issues*

The choice of a spatial weighting and connectivity matrix is a critical step because it can greatly influence the results of spatial analysis. As discussed in the introductory chapters, connectivity (weight) matrices can be obtained from the state of the system using supervisory control and data acquisition (SCADA) and integrated to data fusion architectures.

3.5 Dimensionality reduction

Spatio-temporal models of the form (3.1) lead naturally to the notion of nonlinear spectral dimensionality reduction [27, 28]. In recent years, a large number of methods for dynamic reduction have been proposed. These include linear methods (PCA, linear discriminating analysis, multidimensional scaling) and nonlinear methods (ISOMAP, local linear embedding, diffusion maps). Nonlinear methods offer the advantage of preserving local geometry while achieving dimension reduction and are of interest here.

Given a measurement matrix, **X**, the problem of dimensionality reduction involves determining a *feature* matrix, **D**, that aims to capture or retain certain properties of the data. In physical terms, these techniques map high-dimensional data $\mathbf{x} = \{x_1, x_2, \ldots, x_n\}$ into a lower dimension (a subspace) $\mathbf{y} = \{y_1, y_2, \ldots, y_d\}$ with dimensionality $d < n$, while preserving the geometry of the data as much as possible.

The feature matrix is constructed such that [33, 34]:

1. **D** is square, and its overall size is *mxm*
2. **D** is symmetric since $D_{ij} = D_{ji}$
3. **D** is positive semi-definite, that is, $\mathbf{u}^T\mathbf{D}\mathbf{u} \geq 0$, for all $\mathbf{u} \in \Re^D$

Examples of these matrices include the covariance matrix and distance matrices.

3.5.1 Proximity (similarity) measures

Among the feature matrices, distance matrices are of special physical interest since they can be used to measure dissimilarity. Given a data matrix **X**, the proximity or similarity between objects (trajectories) is described as [35]

$$\mathbf{D} = [d_{ij}] = \begin{bmatrix} d_{11} & \cdots & d_{1m} \\ \vdots & \ddots & \vdots \\ d_{21} & \cdots & d_{mm} \end{bmatrix}, \quad \textit{asymmetric matrix} \tag{3.34}$$

where the coefficients d_{ij} provide a measure of similarity between trajectories.

To pursue this idea further, consider a network of m sensors nonuniformly distributed over the system. Let now the time evolution of a given sensor be given by $x_k(t_j)$, $k = 1, \ldots, m$, $j = 1, \ldots, N$ as illustrated in Figure 3.4.

Let the distance (similarity measure) between two time trajectories $\mathbf{x}_i(t)$, $\mathbf{x}_j(t)$, $t_k = 1, \ldots, N$, up to time t be defined as

$$d_{ij}(t) = \|\mathbf{x}_i - \mathbf{x}_j\|^2 = \sum_{k=1}^{m}(x_{ki} - x_{kj}) \tag{3.35}$$

Generalizations to this model to define instantaneous distances between trajectories are described in Chapter 6.

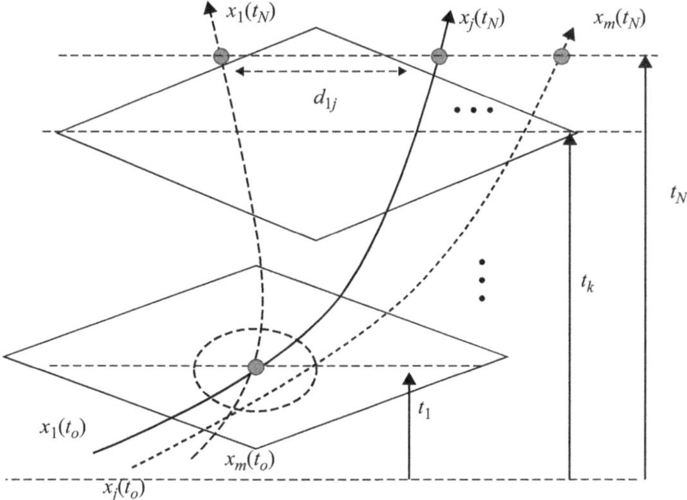

Figure 3.4 Dynamic trajectories for spectral dimensionality reduction

In practice, the pairwise distances (3.35) are combined with a Gaussian kernel of bandwidth ε to retain small *rms* values. This transformation yields the m-by-m matrix **K** (a transition probability kernel), with elements [38]

$$K_{ij} = \exp\left(\frac{-d_{ij}^2}{2\varepsilon}\right) \quad i,j = 1, ..., m \tag{3.36}$$

where $\varepsilon > 0$ represents the local scale; the thresholding operation has the effect of retaining only short pairwise distance. Other choices of the Gaussian kernel are given in [34].

The pairwise affinity matrix can be interpreted as a weighted graph where the measurement points (sensors) are the graph nodes and the weight of the edge connecting node i to node j is the distance d_{ij}. The kernel is symmetric and positivity preserving.

The mxm-dimensional matrix of distances can now be defined as

$$[\mathbf{K}] = [K_{ij}] = \begin{cases} 0 & \text{for } i = j \\ K_{ij} & \text{for } i \neq j \end{cases}$$

where the diagonal elements K_{ii} are zero by definition.

The distance matrix **K** is symmetric and positivity preserving but is not positive semi-definite; this prevents the direct application of spectral dimensionality reduction techniques.

3.5.2 Nonlinear spectral dimensionality reduction

Nonlinear spectral dimensionality reduction techniques seek to alleviate this problem by modeling the data and include dimensionality reduction, clustering, and

Table 3.1 An overview of nonlinear dimensionality reduction methods

Model	Method/Optimization
PCA/POD	Full spectral, Euclidean distance
Isometric feature mapping	Spectral, Geodesic distance
Maximum variance unfolding	Full spectral, kernel-based
Laplacian eigenmaps	Sparse spectral, neighborhood graph Laplacian
Locally linear embedding	Sparse spectral, reconstruction weigths
Diffusion maps	Diffusion distance

data parameterization. These models can be roughly divided into three main categories: spectral graph cuts, eigenmaps, and diffusion maps.

Table 3.1 summarizes the main characteristics of some nonlinear approaches. See [35, 36] for a review of these techniques.

3.5.2.1 Diffusion maps

One way to define a meaningful measure of dynamical proximity between different trajectories is through the use of diffusion maps [37]. Given a distance matrix **X** in (3.1), the diffusion distance can be found by inducing a random walk on the dataset **X** to ensure that the distance matrix is positive definite.

A positive-definite kernel **K** can now be obtained, whose (*i,j*)th element is given by (3.16). Given a matrix, the elements of **K** may be defined in such a way that the transition probabilities p_{ij} from i to j can now be obtained as $p_{ij} = K_{ij}/\sum_{k=1}^{m} K_{ik}$.

Define now a diagonal matrix **D** whose entries are the row sums of **K**, that is

$$\hat{\mathbf{D}} = \begin{bmatrix} \sum_{j=1}^{m} K_{1j} & & & \\ & \sum_{j=1}^{m} K_{2j} & & \\ & & \ddots & \\ & & & \sum_{j=1}^{m} K_{mj} \end{bmatrix}$$

The Markov transition matrix **M** can now be defined as

$$\mathbf{M} = \mathbf{A}^{-1}\mathbf{D} \tag{3.37}$$

with elements $M_{ij} = K_{ij}/\sum_{j=1}^{m} K_{ij},\ i,j = 1,\ldots,m.$

Associated with this matrix is the normalized graph Laplacian matrix $\mathbf{L} = \mathbf{D}^{-1}\mathbf{K} - \mathbf{I}$, where **I** is the *m*-by-*m* identity matrix.

Properties of the Markov transition matrix are summarized as follows:

1. The matrix **M** is nonnegative, unsymmetrical, and row-stochastic (rows sum to 1). It can be shown that the eigenvalues of a stochastic matrix are nonnegative and the largest eigenvalue is 1.
2. Matrix **M** is invariant to the observation modality and is resilient to measurement noise.

The eigenvalue problem for the operator $\mathbf{M}$ can be defined as $\mathbf{M}\psi_j = \lambda_j\psi_j$ with corresponding left eigenvectors φ_j; matrix $\mathbf{M}$ has a complete set of eigenvalues λ_i, of decreasing order of magnitude

$$\lambda_o > \lambda_1 > \ldots > \lambda_{m-1} > 0$$

with $\lambda_o = 1$, and $\psi_o = [1 \quad 1 \quad 1 \quad 1]^T$.

Following [36, 37], let now $\mathbf{M}_s = \mathbf{D}^{1/2}\mathbf{K}\mathbf{D}^{-1/2} = \mathbf{D}^{-1/2}\mathbf{M}\mathbf{D}^{-1/2}$ be a normalized affinity matrix (a normalized kernel) that shares its eigenvalues with the normalized graph-Laplacian $\mathbf{L}$.

Matrix $\mathbf{M}_s$ is symmetric (and therefore diagonalizable) and positive definite with a decomposition $\mathbf{M}_s = \mathbf{U}\Lambda\mathbf{U}^T$, where $\Lambda = diag\{\lambda_o, \lambda_1, \cdots, \lambda_{m-1}\}$ and has a complete set of eigenvectors $\mathbf{U}_j$, $j = 1, \ldots, m-1$, and $\mathbf{U}\mathbf{U}^T = \mathbf{U}^T\mathbf{U} = \mathbf{I}$.

It follows readily that

$$\mathbf{M}_s = \mathbf{D}^{1/2}\mathbf{M}\mathbf{D}^{-1/2} = \mathbf{U}\Lambda\mathbf{U}^T$$

and

$$\mathbf{M} = \underbrace{\mathbf{D}^{-1/2}\mathbf{U}}_{\Psi}\,\Lambda\underbrace{\mathbf{U}^T\mathbf{D}^{1/2}}_{\Phi} = \Psi\Lambda\Phi$$

where

$$\begin{cases} \Psi = \mathbf{D}^{-1/2}\mathbf{U} \\ \Phi = \mathbf{U}^T\mathbf{D}^{1/2} \end{cases}$$

with $\mathbf{U} = [\mathbf{u}_o \quad \mathbf{u}_2 \quad \cdots \quad \mathbf{u}_{m-1}]$.

Therefore the left and right eigenvectors of $\mathbf{M}$ are related to those of $\mathbf{M}_s$ according to [38]

$$\varphi_j = \mathbf{u}_j\sqrt{\mathbf{D}}\,; \quad \psi_j = \mathbf{u}_j 1/\sqrt{\mathbf{D}}$$

The diffusion distance can now be defined in terms of the forward probabilities $\mathbf{P}$ as

$$D_{ij} = \sum_{r=1}^{L} \frac{(M_{ir} - M_{jr})}{\Psi(x_r)}$$

with

$$\Psi(x_m) = \frac{\sum\limits_{j=1}^{L} M_{jm}}{\sum\limits_{k=1}^{L}\sum\limits_{j=1}^{L} M_{jk}}$$

As discussed in [38], (3.14) describes the evolution of a discrete-time diffusion process. The d-dimensional diffusion map is defined at time t, as the map

$$\boldsymbol{\Psi} = [\boldsymbol{\psi}_1 \quad \boldsymbol{\psi}_2 \quad \cdots \quad \boldsymbol{\psi}_{m-1}] = [\boldsymbol{\lambda}_1\boldsymbol{\Phi}_1 \quad \boldsymbol{\lambda}_2\boldsymbol{\Phi}_2 \quad \cdots \quad \boldsymbol{\lambda}_d\boldsymbol{\Phi}_d]^T \tag{3.38}$$

with $\boldsymbol{\psi}_j^T\boldsymbol{\psi}_k = 1, \text{if } j = k, \text{and } \boldsymbol{\psi}_j^T\boldsymbol{\psi}_k = 0, \text{if } j \neq k.$

Based upon the above properties, the eigenvectors $\boldsymbol{\Phi}_i$ can be interpreted as follows: The first eigenvector $\boldsymbol{\Phi}_o$, captures the slowest dynamics (the global trend). The second and subsequent eigenvectors capture the oscillatory dynamics.

The approach offers two major advantages over linear dimensionality reduction methods: diffusion maps are nonlinear, and they preserve local structures. Conceptually, diffusion analysis involves two related steps: dimension reduction and feature extraction that make the procedure especially useful for analyzing power system data.

Techniques to analyze and retrieve diffusion coordinates are discussed next.

3.5.2.2 Time series interpretation of diffusion maps

Following the discussion of the previous sections, each column of the $\mathbf{x}_j$ of the observation matrix can be expressed as

$$\mathbf{x}_j = \sum_{i=1}^{d} c_i \boldsymbol{\Psi}_i$$

where p_d is the unknown intrinsic (space) dimensionality, and the c_i are the time-varying parameters that must be determined.

In analogy with POD analysis in previous sections, the transition probabilities (eigenbehavior) can then be projected onto the physical (data) space through the following transformation

$$a_{j-1}(t) = \mathbf{X}\boldsymbol{\Psi}_j, \quad j = 1, ..., d \tag{3.39}$$

where $a_o(t)$ provides a good approximation to the ensemble mean, and the $a_i(t)$, $i = 1, \ldots, m - 1$ capture the time evolution of the diffusion coordinates.

They are referred to here as the amplitudes or weighting coefficients of the diffusion coordinates and play a similar role to that of the time-dependent coefficients in (3.15).

The group centroid of the time-varying amplitudes can then be computed as

$$\overline{a}(t) = \frac{1}{n}\sum_{j=1}^{n} a_j(t)$$

where n can represent a subset of dynamic trajectories or be equal to the model dimensionality.

A similar interpretation is also possible for the diffusion coordinates. In practice, the dimensionality of the embeddings can be determined from the spectral gap of eigenvalues or using energy criteria as discussed in the numerical implementation of the procedure.

Conceptually, the diffusion process transforms the time series from multiple measurement points into a single time series, preserving as much of the relevant information as possible during the dimensionality reduction.

3.5.2.3 Other approaches

There are other types of nonlinear dimensionality reduction techniques that might be used for feature extraction. Table 3.1 summarizes the main characteristics of some of these approaches [37–39].

Subsequent sections discuss the application of these techniques in the context of modal analysis of power system oscillatory stability.

3.5.2.4 Grouping trajectories

Once the eigenvectors or diffusion coordinates have been computed, the eigenvectors of the transition probabilities can be used to form groups and patterns in which the dominant modes have similar structural features. Clustering techniques based on diffusion coordinates are introduced later in this book [39].

Application of these concepts will now be made to a simple example.

3.6 Motivational example

As a motivational example, the application of space–time models is demonstrated on simulated data of a 5-machine, 10-bus test system adapted from [40]. A single-line diagram of the test system is shown in Figure 3.5. Detailed steady-state and dynamic data for this system is contained in Appendix B.

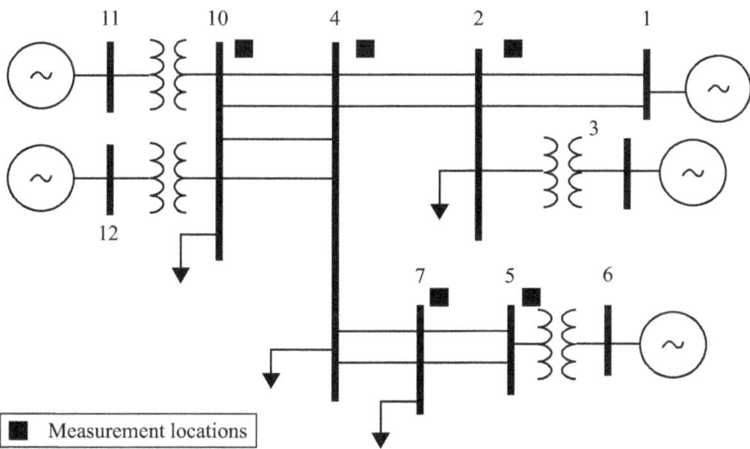

Figure 3.5 Ten-bus, 5-machine test system. Measurement locations are indicated by filled squares

The generator at bus #1 has a large inertia and virtually behaves as an infinite bus. Other generators are represented by a detailed two-axis transient model and equipped with a voltage regulator. Loads were represented by constant impedance characteristics.

3.6.1 Small-signal response

To verify the ability of the method to determine dynamic patterns and investigate the temporal variability of different system features, temporal eigenfunctions were examined.

For reference, the system inter-area modes were first determined using a small signal stability program; the three slowest oscillation modes having frequencies smaller than 1.5 Hz are shown in Table 3.2.

The first mode, in Table 3.2, is seen as an unstable oscillation involving the generators at buses 3, 6, 11, and 12 swinging against the infinite bus. The second mode, on the other hand, represents an oscillation in which generators 3, 11, and 12 swing against the generator 6.

Finally, mode 3 is interpreted as an oscillation involving machines 3 and 11 swinging against machines 6 and 12. Discussion will be restricted to the first two modes.

3.6.2 Large system response

In the analysis that follows, a three-phase stub fault was applied at bus 2 at 0.10 s into the simulation to excite electromechanical modes 1 and 2 cleared by removing the fault at 0.15 s. At 20 s scenario, the signal was sampled at a rate of 30 samples per second.

A 5×3000 matrix of observations was then created from model simulations of voltage time series at buses 2, 4, 5, 7, and 10. The matrix is defined as $\mathbf{X} = [\mathbf{V}_2(t)\ \mathbf{V}_4(t)\ \mathbf{V}_5(t)\ \mathbf{V}_7(t)\ \mathbf{V}_{10}(t)]$, where $\mathbf{V}_j(t), j = 1, \ldots, 5$ is a time vector of bus voltage deviations defined as $\mathbf{V}i(t) = [V_j(t_1)\ V_j(t_2) \ldots\ V_j(t_N)]^T$.

Figure 3.6a shows the time evolution of bus voltage magnitudes following the above disturbance. Figure 3.6b shows the detrended fluctuations.

Careful observation of swing curves in Figure 3.6 shows a nearly oscillatory behavior associated with the slowest oscillation mode. As shown, buses 10 and 5 are seen to swing in opposition to the bus voltages at buses 2, 4, and 7. Large voltage magnitude variations are observed at the intermediate substations at buses 5 and 7.

Table 3.2 Electromechanical modes of the system

Mode	Eigenvalue	Frequency (Hz)	Damping (%)	Swing pattern
1	$0.0026 \pm j3.198$	0.509	−0.080	GENs 3, 6, 11, 12 vs. GEN 1
2	$-0.1399 \pm j5.696$	0.906	2.46	GENs 3, 11, 12 vs. GEN 6
3	$-0.2670 \pm j9.411$	1.497	2.84	GENs 3, 11 vs. GENs 6, 12

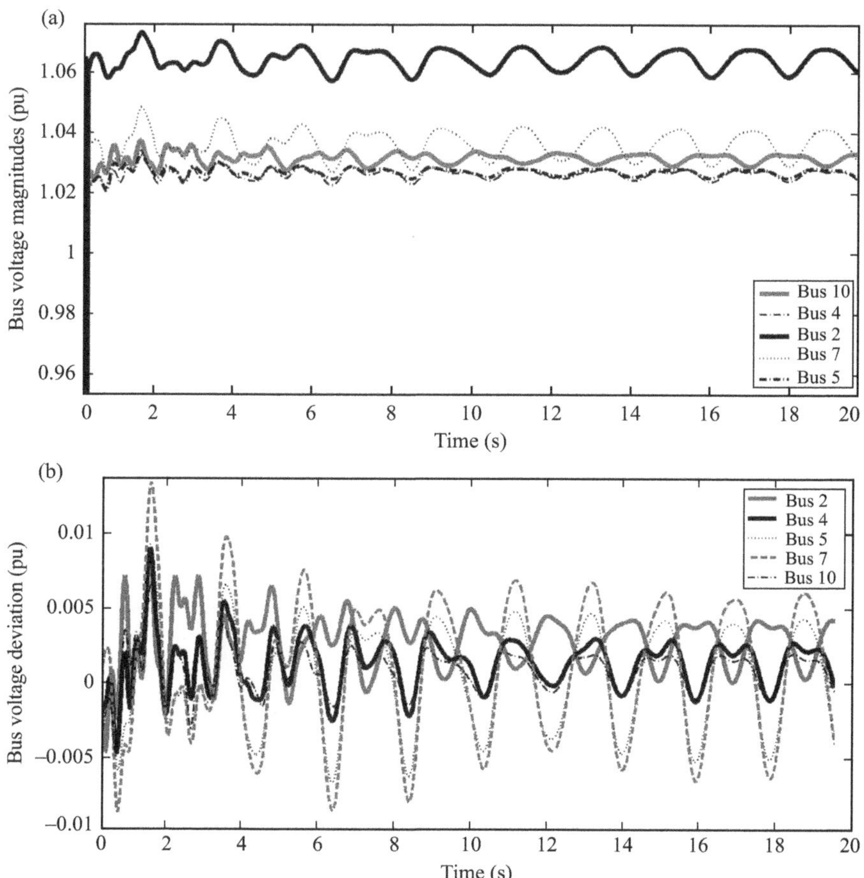

Figure 3.6 Bus voltage magnitudes at load buses: (a) bus voltage deviations at critical load buses; (b) detrended bus voltage magnitudes

3.6.3 Statistical analysis

POD analysis was performed on the observation matrix **X**. Following the above development, the time series can be represented by a linear combination of the eigenfuctions φ_K, as

$$\mathbf{x}_i(t_j) = a_o \varphi_o(\mathbf{x}) + \sum_{k=1}^{p} a_k(t) \varphi_k(\mathbf{x}), \quad i = 1, \ldots, 5 \tag{3.40}$$

where a_o represents the mean process. The data were first detrended by subtraction of the temporal mean.

Figure 3.7a shows the dominant ($i = 1$) temporal and spatial coefficients extracted from (3.6). As depicted in this plot, the temporal eigenfuction exhibits a dominant mode at about 0.5 Hz associated with inter-area mode 1; a second peak at about 0.90 Hz is associated with inter-area mode 2. The associated spectra shown in the inset plot confirms these observations.

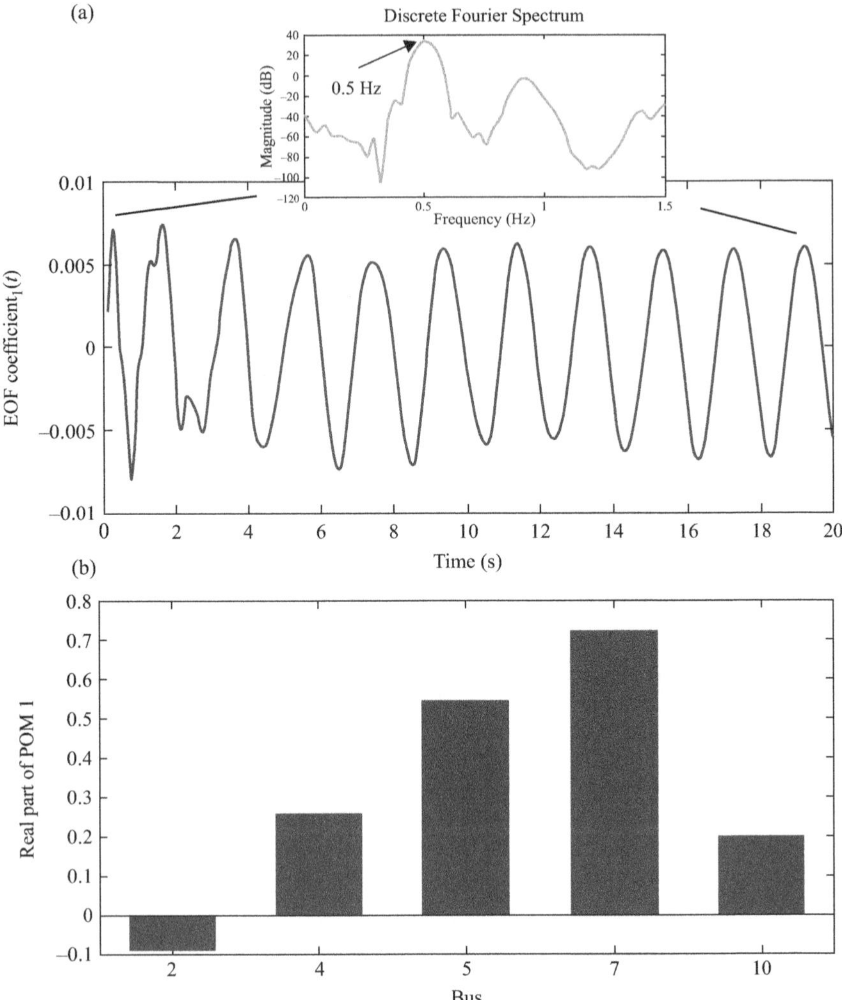

Figure 3.7 Temporal and spatial coefficients associated with the dominant mode at 0.5 Hz for a three-phase fault at bus #2: (a) temporal coefficient; (b) real part of the spatial coefficient, φ_1

From a comparison of the detrended bus voltage deviations in Figure 3.6b, and the spatial coefficient or POM, φ_1, in Figure 3.7b, it is apparent that the technique works well in isolating the dominant pattern of system behavior. As indicated by the real part of the dominant mode in this plots, the bus voltage deviation at bus 2 swings 180° out of phase with the bus voltage deviations at buses 4, 5, 7, and 10. Also of interest, buses 5 and 7 are seen to exhibit the largest voltage deviations in agreement with the voltages traces in Figure 3.7b.

Table 3.3 Prony analysis of the temporal coefficient $a_1(t)$ in Figure 3.7a

Mode	Relative energy	Amplitude	Frequency (Hz)	Damping ratio (%)
1	1.0000	0.00600	0.5102	−0.088
2	0.0094	0.00121	0.9175	2.554
3	0.0004	0.00026	1.4304	2.676

The nature of the POMs becomes evident from Prony analysis of the temporal coefficients in Table 3.3. POM #1 is seen to correspond to inter-area mode 1 in Table 3.2. Interestingly, the damping ratios of the reduced basis are also in good agreements with results in Table 3.2.

3.6.3.1 Spatial prediction

Based on the above model, several studies were conducted to assess the ability of the method to predict system behavior at unmonitored locations. Two simple interpolation methods were evaluated:

1. An inverse distance-weighting (IDW) interpolation $x_i(t) = \sum_{j=1}^{m} w_{ij}x_j(t) \Big/ \sum_{j=1}^{m} w_{ij}$, where all terms are defined as before
2. A connectivity-based interpolation

Numerical results comparing the distance-weighting and connectivity based interpolation techniques are presented below. For completeness, the techniques were applied to voltage and frequency traces.

Figure 3.8 provides a comparison of the full system solution (transient stability) with the solution from the interpolation methods at bus 7. These approaches were compared with simple arithmetic mean estimates from directly measured neighborhood level variables.

For prediction purposes, the evolution of a single unmonitored bus was estimated using information from neighboring modes. As shown in this plot, inverse distance-weighted interpolation methods provide an acceptable estimate of the signal and are adopted for analysis.

The agreement is good though some differences are noted, especially for bus 3 close to the infinite bus for both voltage and frequency traces.

3.7 Sensor placement

The success of WAMS depends critically on the distribution of system sensors. Many methods exist for placing system sensors. A less studied problem is that of determining a small number of observing locations or sensors for capturing specific system behavior (i.e., dominant oscillatory modes) and reconstructing observed measurements. This issue is addressed below.

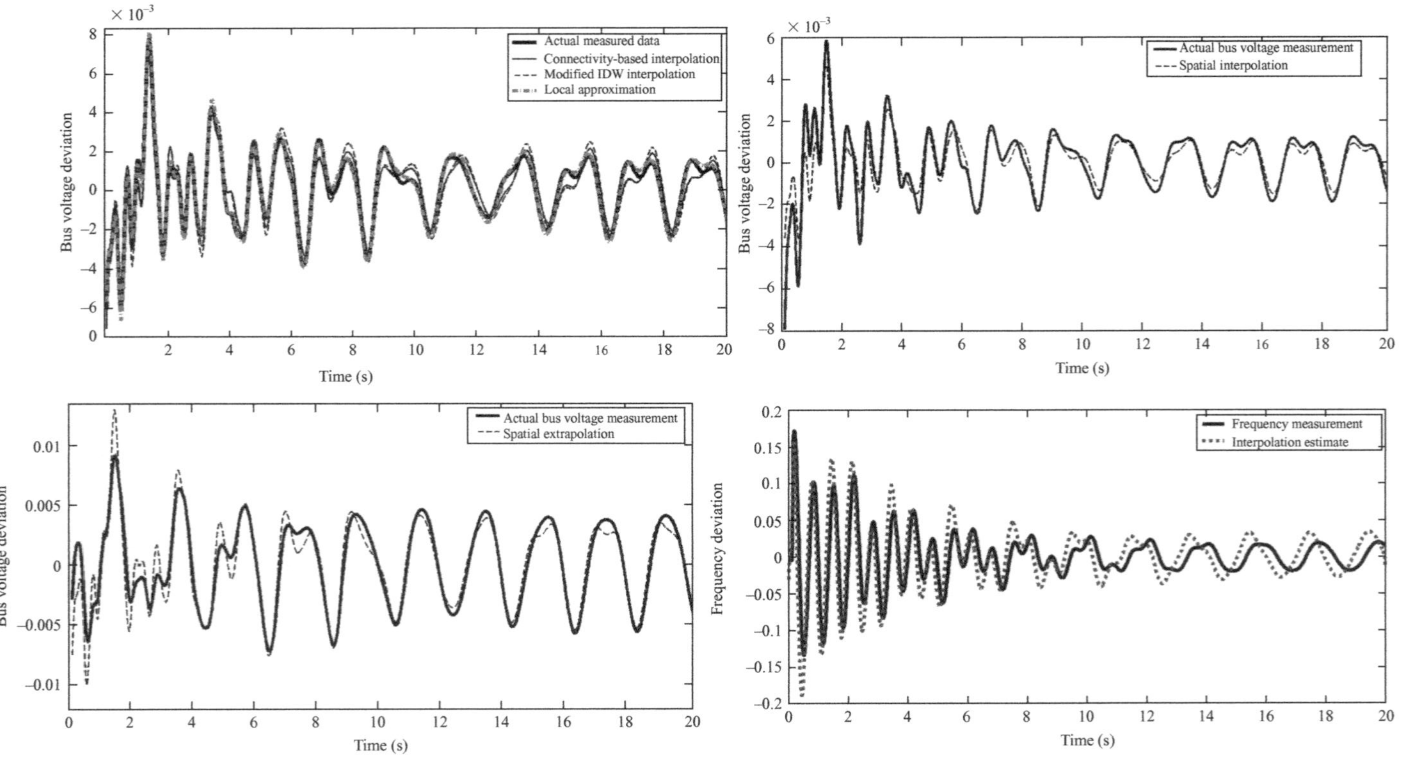

Figure 3.8 Comparison of interpolation estimates and simulated system response: (a) bus 4; (b) actual bus voltage swing plotted against interpolated values of voltage (bus 1); (c) bus 5; (d) bus 1

3.7.1 Problem formulation

The problem of PMU placement and state reconstruction can be posed as a constrained optimization problem using energy criteria. A formal definition of the energy concept is as follows:

Definition 3.1: [41]. Consider a $n \times n$ dimensional matrix **A.** The value $E(\mathbf{A}) = \sigma_1(\mathbf{A}) + \ldots + \sigma_n(\mathbf{A})$ is called the energy of **A**, where the σ_j, $j = 1, \ldots, n$ are the singular values of the matrix.

With this definition of energy in hand, consider a network of m sensors distributed irregularly throughout the system. Following the approach in section 3.3, at any time instant t, all observations can be expressed by an N-by-1 column vector $\mathbf{x}_j = [x_1(t)\ x_2(t), \ldots, x_m(t)]$, $t = t_o, \ldots, N$. The observation matrix is then defined as $\mathbf{X} = [\mathbf{x}_1, \mathbf{x}_2, \ldots, \mathbf{x}_m]$.

In this context, the problem of selection of measurement locations involves solving two related problems:

1. Optimizing sensor locations
2. Optimizing state reconstruction

Insight into these problems can be obtained from the distance matrix (3.29). As pointed out earlier, this equation can be considered as a connected graph in which distances between edges i and j are given by d_{ij}. A simple approach to selecting observing location for capturing the leading modes of variability is choosing the measurement site with least distance with other locations, that is, the product of distances along a path in the graph relates to the total length in the path.

For a single sensor, a good intuitive location is given by the sensor location with the shortest distance d_{ii} with other sensor locations, that is, min d_{ii}.

For the example in section 3.6, the pairwise adjacency matrix is given in Figure 3.9. Also shown in this plot is the connected graph of the system in Figure 3.5.

Table 3.4 depicts the extracted energy from the bus voltage measurements computed using (3.23). Bus 7 is shown to capture the most energy, followed by buses 5 and 4. Table 3.5, in turn, shows the energy captured by simultaneous bus voltage measurements. Here, the first column represents candidate locations for optimal PMU placement.

As suggested, an exhaustive search for all possible combinations of sensors, however, is a hard problem. Selecting l locations from m measurement points involves analyzing $m!/(m-l)!l!$ combinations. This search rapidly becomes infeasible when m and l increase.

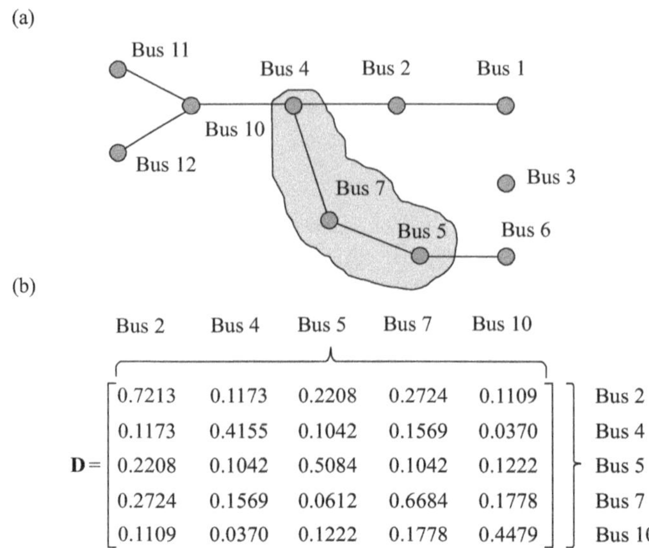

Figure 3.9 Connected graph and distance matrix for the system in Figure 3.5 showing the dominant spatial eigenvector: (a) graph; (b) distance matrix

Table 3.4 Energy contained in the bus voltage measurements

	Bus 2	**Bus 4**	**Bus 5**	**Bus 7**	**Bus 10**
Energy	0.0953	0.0890	0.1695	0.2283	0.0722

Table 3.5 The Cumulative energy of candidate sets

Candidate sets	**Cumulative energy**
Buses 5, 7	0.284
Buses 2, 5, 7	0.286
Buses 2, 4, 5, 7	0.295
Buses 4, 10	0.112

3.7.2 Constrained sensor placement

In practice, measurement points can be selected sequentially. When the problem of interest is wide area monitoring of critical oscillatory modes, the measurement points can be selected to capture the dominant system modes and to reconstruct the observed data [42–44].

In this case, the problem of sensor placement can be naturally cast as a POD-based constrained sensor placement problem from wide-area measurements. Thus, for instance, placement should be optimized to redundant measurements or to take into account measurement errors.

Consider a set of data $\mathbf{X}_l = \{\mathbf{x}_j\}, j = 1, \ldots, N$, collected from synchrophasors, where $\mathbf{x}_j = [x_1(t)\ x_2(t), \ldots, x_m(t)]$. Following Alonso *et al.* [42], given an integer $k(k < m)$, define the kth dimensional set $\mathbf{S}_k = \{\varphi_j\}$, $j = 1, \ldots, k$, as the set of k orthonormal vectors on which the average projection of the data matrix, $\mathbf{X}_l$, is maximized, that is, the k-set that minimizes the average distance to the data.

To illustrate these ideas, let the covariance matrix of $\mathbf{X}_l$ be expressed as

$$\mathbf{R}_X = E\left[\mathbf{X}\mathbf{X}^T\right] \tag{3.41}$$

Let now the eigenvalues and eigenvectors of $\mathbf{R}_\mathbf{X}$ be denoted by $\lambda_1, \lambda_2, \ldots, \lambda_m$ and $\mathbf{U}_1, \mathbf{U}_2, \ldots, \mathbf{U}_m$, respectively. The eigenvectors $\mathbf{U}_j$ are hereafter referred to as the POD modes or *spatial* modes (POMs).

Expanding the measurement matrix, $\mathbf{X}$, onto the POD modes results in

$$\mathbf{X} = \sum_{j=1}^{m} c_j(t)\mathbf{U}_j \tag{3.42}$$

Multiplying (3.42) on the left by $\mathbf{U}_j^T$, it can be easily proved that

$$c_j = \mathbf{U}_j^T\mathbf{X}, \ \ j = 1, \ldots, m$$

where use has made of the biorthogonality properties

$$\begin{cases} \mathbf{U}_i^T\mathbf{U}_i = 1 \\ \mathbf{U}_i^T\mathbf{U}_j = 0 \quad \text{for } i \neq j \end{cases}$$

A number of features are worth noting. First, the variance of c_j equals λ_j, that is, $E[c_j^2] = \lambda_j$. Second, the modal amplitudes are uncorrelated, $E[c_i c_j] = 0$, for $i \neq j$. With these assumptions, it follows immediately that $\mathbf{X} = \mathbf{Uc}$, where $\mathbf{U} = [\mathbf{U}_1, \mathbf{U}_2, \ldots, \mathbf{U}_m]$, and $\mathbf{c} = [c_1, c_2, \ldots, c_m]^T$.

A problem of practical interest is to project the data set $\mathbf{X}$ to itself

$$\mathbf{X}_k = \mathbf{U}_k^T\mathbf{X} \tag{3.43}$$

where $\mathbf{U}_k$ is an appropriate transformation matrix.

Assume now that the network is sensed at k locations $(k < m)$. The distance between the original set of observations and the approximation obtained from a limited number of sensors is given by squared averaged distance [42]

$$D_{av}^2 = \frac{1}{m}\sum_{j=1}^{m} \mathbf{x}_j^T\mathbf{x}_j - \sum_{j=1}^{k} \lambda_j \tag{3.44}$$

where k represents the number of candidate sensor locations.

Remarks:

- The eigenvalues provide a measure of how close the data are to the reduced space, $\mathbf{S}_k$.
- $D_{av} = 0$ when $k = m$, as it is to be expected from physical considerations.

Several methods exist to identify the low-dimensional subspace including guided search methods [45–47]. For the purposes of illustration, the approach in [43] is adopted.

The problem of optimal siting of PMUs can then be cast as an optimization problem subject to the above constraints [43].

$$\max_{X_m} \min_{j=1,\ldots,k} \lambda_j(\mathbf{RR}^T) \tag{3.45}$$

where $\mathbf{R}$ satisfies the eigenvalue problem

$$\mathbf{R}\boldsymbol{\varphi}_j = \boldsymbol{\lambda}_j\boldsymbol{\varphi}_j$$

3.7.2.1 Numerical considerations

Based on previous work [43], a guided search method is explored here to approximate the solution of problems (3.45) but other approaches are possible.

Define the m-by-n operator $\mathbf{P}_m$ as that which projects any n-dimensional vector $\mathbf{v}$ on m of its n coordinates, namely

$$\mathbf{v}_m = \mathbf{P}_m \mathbf{v}_n$$

The estimation problem becomes that of reconstructing the remaining n-m components of v from the available measurements. As discussed in [42], an estimate $\hat{\mathbf{v}}$ can be obtained from

$$\hat{\mathbf{v}}_m = \mathbf{P}_m \boldsymbol{\Phi} \hat{c} \mathbf{v} + \mathbf{P}_m \varepsilon$$

where $\mathbf{P}_m\varepsilon$ is the projection of the error associated with the low-dimensional set on the subspace of measurements. This leads to the standard minimum least-squares problem

$$\min_{\hat{c}} \left(\hat{\mathbf{v}}_m - \mathbf{P}_m \boldsymbol{\Phi} \hat{c} \mathbf{v}\right) \left(\hat{\mathbf{v}}_m - \mathbf{P}_m \boldsymbol{\Phi} \hat{c} \mathbf{v}\right)$$

A solution to this problem is of the form $\hat{c} = (\mathbf{QQ})^T \mathbf{Q} \mathbf{v}_m$. See [42] for details.

The method is based on the observation that the eigenvalues of the matrix $\Sigma\Sigma^T$ are located inside circles centered at the positions given by the diagonal elements (Gershgoring theorem) with radii satisfying

$$r_i = \sum_{j \neq i}^{n} \left| \mathbf{QQ}^T \right|$$

When the radii are much smaller than the diagonal elements *si*, the optimization problem (3.45) reduces to

$$\max_{P_m} \min(s_1, s_2, ..., s_k) \tag{3.46}$$

When this condition is met, $s_1, \ldots, s_k$ in (3.46) correspond with the diagonal elements of the matrix $\mathbf{QQ}^T = \mathbf{\Phi P}_m \mathbf{P}_m^T \mathbf{\Phi}$. Additional details can be found in [42, 43].

This is a *max min* problem, an algorithm that searches and sorts different combination of elements (snapshots) and its summations to solve the above optimization problem. The outcome of the method is an array that identifies the best places to place PMUs to improve dynamic observability of critical inter-area modes.

Table 3.6 summarizes search results for the 10-bus test system. The analysis identifies bus 5 as the best option to place a PMU for all possible combinations. In this analysis, the fractions of energy captured and lost by the reduced order description are defined respectively as

$$E = \frac{\sum_{i=1}^{k} \lambda_i}{\sum_{i=1}^{m} \lambda_i}$$

and

$$L = 1 - E$$

Table 3.7 shows the energy captured by the proper orthogonal modes. As shown, three modes are seen to capture over 99.8% of the total energy. With these approaches, both optimal placement of PMUs and optimal state reconstruction are achieved.

A more detailed analysis of the case study is presented in [42].

Table 3.6 Energy contained in the selected optimal sequences

Optimal sequence	***L***
Buses 5, 7, 4, 10, 2	4.270
Buses 5, 2, 7, 10, 4	4.291

Table 3.7 Energy captured by the POMs

Eigenvalue	**Cumulative energy captured (%)**
1	92.73
2	98.95
3	99.87
4	99.78
5	100

References [44–47] explore the use of optimization techniques to site PMUs. General guidelines for siting PMUs are provided in [48].

References

1. John F. Hauer, William A. Mittelstadt, Kenneth E. Martin, James W. Burns, Harry Lee, John W. Pierre, Danil J. Trudnowski, 'Use of the WECC WAMS in wide-area probing tests for validation of system performance and modeling', *IEEE Transactions on Power Systems*, vol. 24, no. 1, February 2009, pp. 250–257.
2. Mladen Kezunovic, Ali Abur, 'Merging the temporal and spatial aspects of data and information for improved power system monitoring applications', *Proceedings of the IEEE*, vol. 93, no. 11, November 2005, pp. 1909–1919.
3. A. R. Messina, V. Vittal, 'Extraction of dynamic patterns from wide-area measurements using empirical orthogonal functions', *IEEE Transactions on Power Systems*, vol. 22, no. 2, May 2007, pp. 682–692.
4. Marcelo Godoy Simoes, Robin Roche, Elias Kyriakides, Sid Suryanarayanan, Benjamin Blunier, Kerry D. McBee, Phuong H. Nguyen, ... Abdellatif Miraoui, 'A comparison of smart grid technologies and progresses in Europe and the U.S.', *IEEE Transactions on Industry Applications*, vol. 48, no. 4, July/August 2012, pp. 1154–1162.
5. Yoshihiko Susuki, Igor Mezic, 'Nonlinear Koopman modes and coherency identification of coupled swing dynamics', *IEEE Transactions on Power Systems*, vol. 26, no. 4, November 2011, pp. 1894–1904.
6. Kejun Mei, Steven M. Rovnyak, Chee-Mun Ong, 'Clustering-based dynamic event location using wide-area phasor measurements', *IEEE Transactions on Power Systems*, vol. 23, no. 2, May 2008, pp. 673–679.
7. A. R. Messina, J. Nuno, I. Moreno, 'Monitoring the health of large-scale power systems: A near real-time perspective', 8th IFAC Symposium on Fault Detection, Supervision and Safety of Technical Processes, SafeProcess 2012, August 2012, Mexico.
8. Arturo R. Messina (ed.), Inter-Area Oscillations in Power Systems, *A Nonlinear and Nonstationary Perspective*, Springer, New York, NY, 2009.
9. Daniel Karlsson, Morten Hemmingsson, Sture Lindahl, 'Wide area system monitoring and control', *IEEE Power and Energy Magazine*, September/ October 2004, pp. 68–76.
10. A. R. Messina, Vijay Vittal, 'Extraction of dynamic patterns from wide-area measurements using empirical orthogonal functions', *IEEE Transactions on Power Systems*, vol. 22, no. 2, May 2007, pp. 682–692.
11. John F. Hauer, Navin B. Bhatt, Kirit Shah, Sharma Kolluri, 'Performance of WAMS East in providing dynamic information for the north east blackout of August 14', 2004 IEEE Power Engineering Society General Meeting, June 2004, Denver, CO.

12. A. R. Messina, Vijay Vittal., 'Extraction of dynamic patterns from wide-area measurements using empirical orthogonal functions', *IEEE Transactions on Power Systems*, vol. 22, no. 2, May 2007, p. 682.
13. Mladen Kezunovic, Sakis Meliopoulos, Vaithianathan Venkatasubramanian, Vijay Vittal, *Applications of Time-Synchronized Measurements in Power Transmission Networks*, Power Electronics and Power Systems Series, Springer, Cham, Switzerland, 2014.
14. Junshan Zhang, Vijay Vittal, Peter Sauer, 'Networked information gathering and fusion of PMU data – Future grid initiative white paper', Power Systems Engineering Research Center, PSERC Publication 12-07, May 2012.
15. Andrew R. Solow, 'Statistics in atmospheric science', *Statistical Science*, vol. 18, no. 4, 2003, pp. 422–429.
16. Troy R. Smith, Jeff Moehlis, Philip Holmes, 'Low-dimensional modeling of turbulence using the proper orthogonal decomposition: A tutorial', *Nonlinear Dynamics*, vol. 41, 2005, pp. 275–307.
17. A. Hannachi, I. T. Jollife, D. B. Stephenson, 'Empirical orthogonal functions and related techniques in atmospheric science: A review', *International Journal of Climatology*, vol. 27, 2007, pp. 1119–1152.
18. John E. Kutzbach, 'Empirical eigenvectors of sea-level pressure, surface temperature and precipitation complex over North America', *Journal of Applied Meteorology*, vol. 6, October 1967, pp. 791–802.
19. A. R. Messina, P. Esquivel, F. Lezama, 'Time-dependent statistical analysis of wide-area time-synchronized data', *Mathematical Problems in Engineering*, vol. 2010, 2010, pp. 1–17.
20. Alberto Alvarez, 'Performance of satellite-based ocean forecasting (SOFT) systems: A study in the Adriatic sea', *Journal of Atmospheric and Oceanic Technology*, vol. 20, May 2003, pp. 717–729.
21. Y. C. Liang, H. P. Lee, S. P. Lim, W. Z. Lin, K. H. Lee, C. G. Wu, 'Proper orthogonal decomposition and its applications – Part I: Theory', *Journal of Sound and Vibration*, vol. 252, no. 3, 2002, pp. 527–544.
22. Gaetan Kerschen, Jean-Claude Golinval, Alexander F. Vakakis, Lawrence A. Bergman, 'The method of proper orthogonal decomposition for dynamical characterization and order reduction of mechanical systems: An overview', *Nonlinear Dynamics*, vol. 41, 2005, pp. 147–169.
23. G. Kerschen, J. C. Golinval, 'Physical interpretation of the proper orthogonal modes using the singular value decomposition', *Journal of Sound and Vibration*, vol. 249, no. 5, 2002, pp. 849–865.
24. G. A. Webber, R. A. Handler, L. Sirovich, 'The Karhunen–Lóeve decomposition of minimal channel flow', *Physics of Fluids*, vol. 9, no. 4, April 1997, pp. 1054–1066.
25. Nadine Aubry, Régis Guyonnet, Rcardo Lima, 'Spatiotemporal analysis of complex signals: Theory and applications', *Journal of Statistical Physics*, vol. 64, nos. 3/4, 1991, pp. 683–737.

26. Breda Munoz, Virginia M. Lesser, Fred L. Ramsey, 'Design-based empirical orthogonal function model for environmental monitoring data analysis', *Environmetrics*, vol. 19, 2008, pp. 805–817.
27. W. Luo, M. C. Taylor, S. R. Parker, 'A comparison of spatial interpolation methods to estimate continuous wind speed surfaces using irregularly distributed data from England and Wales', *International Journal of Climatology*, vol. 28, 2008, pp. 947–959.
28. Cort J. Willmott, Kenji Matsuura, 'Smart interpolation of annually air averaged air temperature in the United States', *Journal of Applied Meteorology*, vol. 34, December 1995, pp. 2577–2586.
29. Nina Siu-Ngan Lam, 'Spatial interpolation methods: A review', *The American Cartographer*, vol. 10, no. 2, 1983, pp. 129–149.
30. P. de Jong, C. Sprenger, F. van Veen, 'On extreme values of Moran's I and Geary's c', *Geographical Analysis*, vol. 16, no. 1, January 1984, pp. 17–24.
31. Daniel Borcard, Pierre Legendre, 'All-scale spatial analysis of ecological data by means of principal coordinates of neighbor matrices', *Ecological Modelling*, vol. 153, 2002, pp. 51–68.
32. P. Lagonotte, J. C. Sabonnadiére, J. Y. Léost, J. P. Paul, 'Structural analysis if the electrical system: Application of secondary voltage control in France', *IEEE Transactions on Power Systems*, vol. 4, no. 2, May 1989, pp. 479–486.
33. Stéphane Dray, Pierre Legendre, Pedro R. Peres-Neto, 'Spatial-modelling: A comprehemsive framework for principal coordinate analysis of neighbor matrices (PCNM)', *Ecological Modelling*, vol. 196, 2006, pp. 483–493.
34. Pierre Legendre, Louis Legendre, *Numerical Ecology*, Elsevier Science, Amsterdam, Netherlands, 1998.
35. W. Hardle, L. Simar, *Applied Multivariate Statistical Analysis*, Springer, Heidelberg, Germany, 2007.
36. Yair Weiss, 'Segmentation using eigenvectors: A unifying view', Seventh International Conference on Computer Vision, 1999.
37. Boaz Nadler, Stéphane Lafon, Ronal R. Coifman, 'Diffusion maps, spectral clustering and eigenfunctions of Fokker–Planck operators', *Advances in Neural Information Processing*, vol. 18, 2005, pp. 955–962.
38. Mary A. Rohrdanz, Wenwei Zheng, Mauro Maggioni, Cecilia Clementi, 'Determination of reaction coordinates via locally scaled diffusion map', *Journal of Chemical Physics*, vol. 134, no. 124116, 2011, pp. 124116-1–124116-11.
39. Nasir Rajpoot, Muhammad Arif, Abhir Bharelao, 'Unsupervised learning of shape manifolds', British Machine Vision Conference, 2007
40. *A Study of Reactive Power Compensators for High-Voltage Power Systems, Advanced Systems Technology Division and Transmission and Distribution System Engineering Department*, Westinghouse Electric Corporation, Contract 4-L60-6964P, Final Report, May 1981.
41. R. Balakhrishnan, 'The energy of a graph', *Linear Algebra Applications*, vol. 387, 2004, pp. 287–295.

42. Antonio A. Alonso, Christos E. Frouzakis, Ioannis G. Kevrekidis, 'Optimal sensor placement for state reconstruction of distributed process systems', *Process Systems Engineering*, vol. 50, no. 7, July 2004, pp. 1438–1452.
43. M. A. Pérez G., Noe Reyes, A. R. Messina, 'Sensor placement and optimal state recostruction from wide-area measurements', IEEE Transmission and Distribution Conference, 2014.
44. Paritosh Mokhasi, Dietmar Rempfer, 'Optimized sensor placement for urban flow measurement', *Physics of Fluids*, vol. 16, no. 5, May 2004, pp. 1758–1764.
45. V. Madani, M. Parsashar, J. Giri, S. Durbha, F. Rahmatian, D. Day, M. Adamiak, G. Sheble, 'PMU placement considerations – A roadmap for optimal PMU placement', 2011 IEEE/PES Power Systems Conference and Exposition (PSCE), March 2011, Phoenix, AZ.
46. Innocent Kamwa, Robert Grondin, 'PMU configuration for system dynamic performance measurement in large multiarea power system', *IEEE Transactions on Power Systems*, vol. 17, no. 2, May 2002, pp. 385–394.
47. Aranya Chrakabortty, Clyde F. Martin, 'Optimal measurement allocation algorithms for parametric model identification of power system', *IEEE Transactions on Control Systems Technology*, vol. 22, no. 5, September 2014, pp. 1801–1812.
48. Joe H. Chow (ed.), *Guidelines for Siting Phasor Measurement Units*, North American SynchroPhasor Initiative Research Initiative Task Team (RITT) Report, June 2011.

Chapter 4
Advanced data processing and feature extraction

4.1 Introduction

Power system data is often corrupted by different artifacts and noise that are often non-Gaussian, nonlinear, and nonstationary. High levels of ambient noise, in particular, result in nonstationary signals, which may lead to inefficient performance of conventional data processing methods.

Extracting robust parameters from such signals, and providing confidence in the estimates, is therefore difficult and requires an adaptive filtering approach that accounts for artifact types [1–3].

The extensive development of signal processing methods for measured data during the last decade has been guided by the study of modal properties of dominant inter-area modes. Much of the literature on signal processing focuses on linear analysis. Linear, stationary methods are successful when carefully applied, but they lack the general applicability offered by a data-driven approach.

Traditionally, modeling techniques have dealt with complex behavior by trying to apply linear models to windows of observation exhibiting nearly stationary or linear characteristics.

Under stressed operating conditions it may be possible to parameterize system behavior or even decide on the appropriate form of linear model. Only very recently, techniques that account for nonlinear and nonstationary behavior have begun to percolate into power system data processing theory. In parallel with this, new statistical techniques for identifying trends, quasi-stationary behavior, and other measures of predictability continue to be developed. Explicit treatment of these issues has led to different data processing approaches with the ability to process more general system behavior.

In spite of the prevalence of such a large number of signal processing methods, and its success in a number of practical applications, modal analysis remains a difficult challenge.

In this chapter, nonlinear and/or nonstationary data processing methods are examined. Multivariate analysis methods are also considered, and the concept of feature extraction and selection is introduced. The methods are contrasted to earlier standard analysis approaches in existing literatures. Examples are used throughout to illustrate various points.

4.2 Power oscillation monitoring

There has been an extensive number of research efforts focused around modeling power system oscillatory behavior. Due to the breath of these efforts, emphasis here is placed on the analysis of nonlinear and/or nonstationary methods. An excellent review of recent linear and nonlinear analysis methods for power oscillation monitoring has been presented by Sánchez Gasca *et al.* [2], which also includes a number of application examples.

As noted above, measured data may contain noise, different levels of unavoidable offset, trends, large amounts of data points, etc. [2, 4, 5]. At its most generic level, a measured signal, $x(t)$, can be represented by a general time-varying model of the form

$$x(t) = \underbrace{\mu(t)}_{Trend} + \underbrace{\sum_{j=1}^{p} x_j(t)}_{\substack{Oscillatory \\ components}} + \underbrace{r_p(t)}_{\substack{Residual \\ information}} + \underbrace{\varepsilon(t)}_{Noise} \tag{4.1}$$

where $\mu(t)$ is a low-frequency or trend component, $x_j(t)$ represent dominant oscillatory behavior, $r_p(t)$ is a residual, and $\varepsilon(t)$ represents noise. Other more general representations are possible.

Figure 4.1 illustrates the behavior of typical measured data showing typical components in (4.1). Equation (4.1) is a very general version of various existing linear and nonlinear time domain decomposition methods. For reference, Fourier and Prony models are also included.

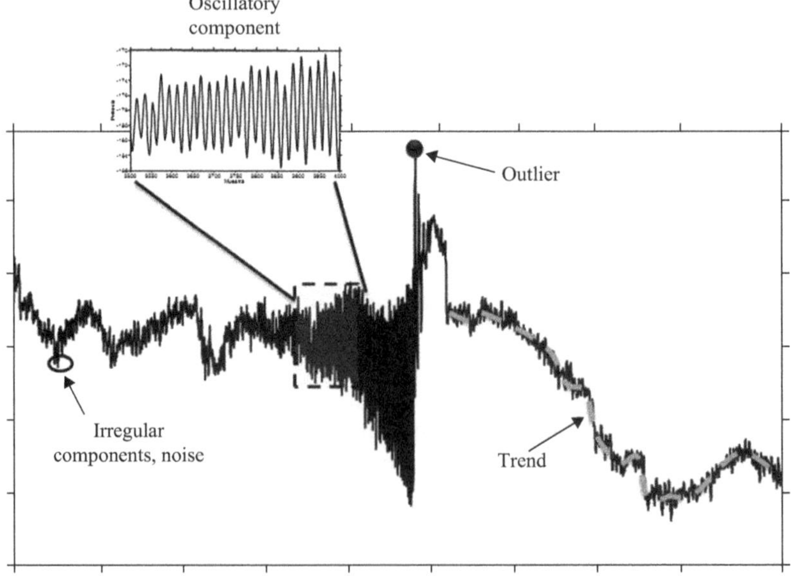

Figure 4.1 Measured signal illustrating the nature of typical observed behavior

Table 4.1 Modal decomposition methods

Method	Model structure
Fourier	$x(t) = \sum_{j=1}^{p_f} A_j \cos(\omega_j t + \theta)$
Prony	$x(t) = \sum_{j=1}^{p_p} A_j e^{-\sigma_j t} \cos(\omega_j t + \theta)$
Dynamic harmonic regression	$x(t) = \mu(t) + \sum_{j=0}^{R} \left[a_j(t)\cos(\omega_j t) + b_j(t)\sin(\omega_j t)\right] + e(t)$
HHT	$x(t) = \sum_{j=1}^{p} c_j(t) + r_k(t)$
Wavelet analysis	$x(t) = \sum_{k=1}^{p_w} d_j(t) + r_k(t)$
Additive models	$x(t) = \sum_{i=1}^{p_a} \beta_o + f_i(t) + r(t)$

Table 4.1 summarizes some particular cases of this model. The reader is referred to [2] for a description of other modal approximations.

In the following sections, a brief introduction to various prominent time-series analysis techniques and their extensions is presented.

4.3 Time-frequency representations

Recently, new analytical techniques with promise of broad applicability have emerged. Among the alternative time-frequency formulations, Hilbert–Huang transform (HHT) and wavelet models have proven to be particularly useful for signal extraction, forecasting, and backcasting of time series.

Despite their different origins, these decompositions are of the general form

$$x(t) = \sum_{j=1}^{p} x_j(t) + r_p(t) \tag{4.2}$$

where $x_j(t)$ represents time-varying oscillatory modes, p denotes the number of modes or time-varying functions, and r_p is a trend.

Various motivating factors lead to this kind of decomposition. First, these decompositions extend linear analysis to the nonlinear and/or nonstationary setting. Second, these representations lead naturally to the notion of nonlinear detrending and filtering.

4.3.1 Hilbert–Huang analysis

The HHT is a multiscale method based on two recently proposed mathematical techniques: empirical mode decomposition (EMD) and the Hilbert transform. For a background on the numerical aspects of the method, see [6–14].

The method assumes that the measured time series can be decomposed in terms of a finite number of harmonic components in the form of "fast" nearly monochromatic oscillations that are modulated by "slow" varying amplitudes. Hence, the method is based on slow–fast partition of the measured dynamics and on the correspondence between analytical and empirical (i.e., derived from EMD of the measured time series) slow-flow models [11].

The essence of this technique is to identify and extract custom time-varying oscillating components from the system response that can be associated with different timescales called *intrinsic mode functions* (IMFs) through a process called *sifting*.

Following the same notation as Messina and Vittal [7] and Huang *et al.* [6], this decomposition can be represented mathematically as

$$x(t) = \sum_{j=1}^{p} x_j = \sum_{j=1}^{p} c_j(t) + r_p(t) \tag{4.3}$$

where subscript j represents the spatial position or sensor, c_j is the jth IMF, p is the number of IMFs, and r_p is the residue.

Physically, the decomposition (4.3) can be rewritten in the more useful form

$$x(t) = \sum_{j=1}^{p} x_j = \sum_{j=1}^{p} A_j(t)\cos\varphi_j(t)$$

where $A_j(t)$ and $\varphi_j(t)$ are the instantaneous amplitude and phase of the jth component, respectively, and

$$\varphi_j(t) = \omega_{c_j}(t) + \int_0^t \omega_j(\tau)d\tau + \theta_j$$

in which ω_j is the carrier angular frequency, $\omega_{c_j}(t)$ is the frequency modulating signal, and θ_j is the phase offset of the component.

Essentially, each IMF is an amplitude-modulated–frequency-modulated (AM/FM) signal satisfying

$$A_j(t) > 0, \quad \omega_j = \frac{d\varphi_j(t)}{dt} > 0, \quad \forall t$$

It has been observed that the change in time in $A_j(t)$ and $d\varphi_j(t)/dt$ is much slower than the change of $\varphi_j(t)$ itself [15].

The IMFs may be thought of as simple, time-varying oscillatory mode with different amplitude and frequency content; by construction, the IMFs are nearly orthogonal. The first IMF captures the highest frequency content; the frequency content decreases with the increase in IMF.

Let $x(t)$ be a real measured signal, and $x_H(t)$ be its Hilbert transform. Given a model of the form (4.3), a complex signal $z(t)$ can be constructed by adding an imaginary signal to the original function

$$z(t) = \sum_{j=1}^{p} x_j(t) + ix_{H_j}(t) = \sum_{j=1}^{p} A_j(t)e^{i\varphi_j(t)} \tag{4.4}$$

where $A_j(t) = \sqrt{x_j^2(t) + x_{H_j}^2(t)}$, $\varphi_j(t) = \arctan(x_{H_j}(t)/x_j(t))$ are the instantaneous amplitude and phase of the local time-varying wave, and $x_H(t) = H[x] = \frac{1}{\pi} PV \int_{-\infty}^{\infty} \frac{x(\tau)}{t-\tau} d\tau$ is the Hilbert transform of $x(t)$; $PV \int_{-\infty}^{\infty}$ denotes the Cauchy principal value, $PV \int_{-\infty}^{\infty} = \lim_{\varepsilon \to 0} \left[\int_{-\infty}^{t-\varepsilon} + \int_{t+\varepsilon}^{+\infty} \right]$.

Using the above representation, each IMF, $c_j(t)$, can be expressed as

$$c_j(t) = \mathrm{Re}\left(A_j(t) e^{i\varphi_j(t)} \right) = A_j(t) \cos \varphi_j(t) \tag{4.5}$$

which describes AM–FM single component signals.

The HHT methodology introduces a couple of novel features: (a) both trends and modal parameters can be determined simultaneously and (b) because of the data-adaptive nature of the base functions, φ_j, the HHT technique allows for the modeling of nonperiodic oscillations.

The original signal can then be expressed as the real part of the complex expansion

$$\begin{aligned} x(t) &= \mathrm{Re} \sum_{j=1}^{p} z_j(t) \\ &= \mathrm{Re}\left[\sum_{j=1}^{p} x_j(t) + iH\left[x_j(t)\right] \right] \\ &= \mathrm{Re}\left[\sum_{j=1}^{p} A_j(\omega, t)\, e^{j \int_o^t \omega_j(t) dt} \right] \end{aligned} \tag{4.6}$$

where the amplitude A_j is a function of ω_j and t, and defines a generalized form of the Fourier spectra with time-varying amplitudes and phases.

4.3.1.1 Empirical mode decomposition

The EMD method introduced by Huang provides an analytical basis for the nonlinear decomposition of a signal $x(t)$ into a finite set of essentially band-limited components or basis functions called IMFs.

As discussed earlier, EMD has its foundations in the notion that any oscillatory signal consists of two parts: (a) a slowly varying trend or residue and (b) a fast component superimposed on the slow component [5, 12].

Distinct from previous methods, the transformation is complete, nearly orthogonal, adaptive, and total (the original signal may be recovered by summing the IMF components). Completeness, in particular, depends on the accuracy of the extraction process. In addition, orthogonality is also critical in isolating and identifying local timescales.

Sifting process

Central to the computation of efficient basis functions is the extraction technique. As highlighted in the previous section, the EMD is based on the simple physical assumption that any signal $x(t)$ consists of the sum of different simple IMFs.

More formally, an IMF is defined as a wave where the following conditions are met:

1. In the whole time span of the signal, the total number of extremes, namely maxima and minima, N_{max} and N_{min}, and the number of zero crossings, N_{zeros}, must be equal or differ at most by 1, that is

$$N_{max} + N_{min} - N_{zeros} = \pm 1$$

2. At any time instant, the mean value of the amplitudes defined by the local maxima e_{max} and minima e_{min} must be zero:

$$(e_{max}(t) + e_{min}(t))/2 = 0$$

Because of physical constraints, the mean value of the IMFs is never zero since this involves the definition of a local timescale.

The IMFs are found using a recursive procedure called *sifting*, which generates the highest frequency IMF first. The basic EMD algorithm to extract the IMFs can be summarized as follows [1, 7]:

Empirical mode decomposition algorithm

Step 1. Starting with the original signal $x(t)$, set $r_o(t) = x(t)$, and $j = 1$

Step 2. Extract the jth IMF using the following iterative sifting procedure:

(a) Set $h_o(t) = r_j(t)$ and $i = 1$

(b) Identify the successive local maxima and the local minima. The time spacing between successive maxima is defined to be the timescale of the successive maxima.

(c) Interpolate the local minima and the local maxima with a cubic spline or other similar techniques. Form an upper envelope $e_{max_i}(t)$ and a lower envelope $e_{min_i}(t)$ for the whole data span.

(d) Compute the instantaneous mean of envelopes $m_{i-1}(t) = (e_{max_i}(t) - e_{min_i}(t))/2$ and subtract it from $h_i(t)$. Determine a new estimate h^i_q, $q = 1, \ldots, n$ using the recursive relations

$$h^i_q(t) = h^i_{q-1}(t) - m^i_q(t), \quad q = 1, \ldots, n_i \tag{4.7}$$

for $i = 1, \ldots, n$, where $h^i_o(t) = r_o(t) = x(t)$, for $i = 1$, such that $e_{min_i}(t) \leq h_i(t) \leq e_{max_i}(t)$ for all t. Set $i = i + 1$.

(e) Repeat the above procedure until $h_i(t)$ satisfies a predetermined stopping criterion. Then, set $c_j(t) = h_i(t)$.

Step 3. Obtain an improved residue $r_j(t) = r_{j-1}(t) - c_j(t)$. Repeat the above steps with $j = j + 1$ until the number of extrema in $r_j(t)$ is less than 2. When successful, the result of this procedure is a residual $r_i(t) = r_{i-1}(t) - c_i(t)$, with $c_i(t) = h^i_{n_i}$, that contains information about higher frequency components.
The residual $r_i(t)$ is then treated as a new signal and the process is repeated for the new signal ($i = i + 1$). The process concludes when there are no longer any maxima or minima in the residual.

The criterion used to stop the sifting is critical to this procedure. In its original formulation, the process of sifting for an IMF stops if the value of the normalized square difference between successive values of $h_q^i(t)$ is smaller than a pre-set value:

$$\mathrm{SD} = \sum_{t=0}^{n} \left[\frac{\left[h_q^i(t) - h_{q-1}^i(t)\right]^2}{(h_{q-1}^i(t))^2} \right] \leq \text{threshold}$$

A critical assessment of the performance of this criterion to satisfy the above condition along with a review of recent extensions is provided in [5]. Reference [6] describes other criteria used in publicly available software.

The sifting process serves mainly two purposes: (i) eliminating riding waves and (ii) making the wave profiles more symmetric. Figure 4.2 gives a schematic representation of the sifting process. Appendix C describes the use of masking techniques to improve the EMD.

Referring to Figure 4.2, the outcome of the sifting process is a decomposition of the form

$$\begin{aligned} h_q^i(t) &= h_{q-1}^i(t) - m_q^i(t) \\ c_q^i(t) &= h_q^i(t) \\ r_i(t) &= x(t) - c_i(t) \\ q &= 1, \ldots, n_i \end{aligned}$$

Upon convergence, the signal $x(t)$ is decomposed into a set of nearly decoupled nonstationary (and possibly nonlinear) modes.

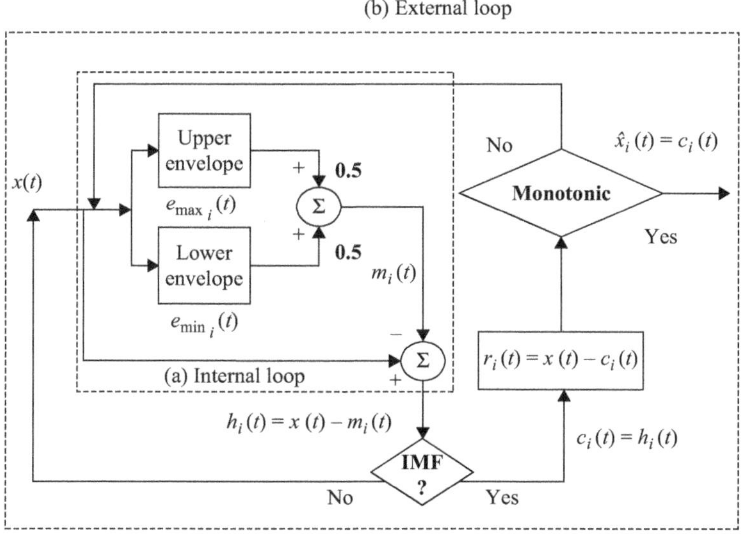

Figure 4.2 Empirical mode decomposition

Link with other decomposition methods

Table 4.2 summarizes related modal decompositions equivalent to the EMD procedure.

Table 4.2 Local signal decomposition methods

Method	Reference
Hilbert vibration decomposition	Feldman [13]
Local mean decomposition	Chen *et al.* [14]
Synchrosqueezed wavelet	Daubechies *et al.* [15]
Modified local mean decomposition	Smith [16]
L_p periodogram	Lauria and Pisani [17]

4.3.1.2 Damping and frequency characterization

Once the signal is decomposed into nearly orthogonal modes, modal parameter can be estimated using the notion of the analytic signal. In light of this idea, assume now that the jth IMF mode analytic signal can be expressed in the form [14]

$$z_j(t) = A_j(t)e^{-\varphi_j(t)+i\theta_{m_j}(t)} = \underbrace{\Lambda_j(t)}_{\substack{Slow\\ component}} \underbrace{e^{-\varphi_j(t)}}_{\substack{Fast\\ component}} \tag{4.8}$$

and

$$\dot{z}_j(t) = -\Lambda_j(t)e^{-\varphi_j(t)}\omega_j(t) + e^{-\varphi_j(t)}\dot{\Lambda}_j(t) \tag{4.9}$$

where $\Lambda_j(t) = A_j(t)e^{i\theta_j(t)}$, $\varphi_j = \int_o^t \sigma_j(t)dt$ is an exponential factor characterizing the time-dependent decay of the waves for the jth components, and $\sigma_j(t)$ is the associated instantaneous damping.

Making use of this assumption in (4.8) and (4.9) results in

$$\frac{\dot{z}_j(t)}{z_j(t)} = \left[\left(\sigma_j(t) + \frac{\dot{A}_j(t)}{A_j(t)}\right) + i\omega_j(t)\right] \tag{4.10}$$

where the overdot indicates differentiation with respect to time.

It can then be readily seen that

$$\sigma_j(t) = \text{Re}\left[\dot{z}_j(t)/z_j(t)\right]$$

and

$$\omega_j(t) = \text{Im}\left[\dot{z}_j(t)/z_j(t)\right]$$

Chapter 8 explores these concepts in further details.

4.3.1.3 Phase characterization

Phase information is of special relevance to the analysis of measured data. Physically, the instantaneous frequency, ω, can be interpreted as the rate of change of phase with respect to time as [3]

$$\omega(t_k) = \frac{\varphi(t_k) - \varphi(t_{k-1})}{t_k - t_{k-1}} \tag{4.11}$$

In the limit $\Delta t = t_k - t_{k-1} \to 0$, (4.11) defines the instantaneous frequency.

To obtain the phase evolution of the oscillation, the magnitude of the analytic signal is projected onto the unit circle.

$$\varphi(t_k) = \frac{z}{\|z\|} = e^{i\varphi(t_k)}$$

This equation describes a time-varying phasor in the Cartesian plane.

Recently, Senroy presented a theoretical basis for Hilbert analysis of measured data that incorporates phase information [18]. Let the instantaneous phase of signal i be φ_i, and that of signal j be φ_j. An instantaneous coherency index for two signals may be defined as the difference between their instantaneous phase angles.

$$\left|\varphi_i(t) - \varphi_j(t)\right| = 2n\pi \tag{4.12}$$

where $n = 0, 1, \ldots.$

The value of this index indicates the degree of coherency between the two signals. For modal analysis, the components i, j should correspond to fully decoupled modes.

4.3.1.4 Detrending of nonlinear and nonstationary time series

As seen in the earlier derivation, EMD offers a natural way to extract from the original signal an instantaneous time-varying trend. From (4.7), it is readily apparent that

$$\begin{cases} r_1(t) = x(t) - h^1_{n_1} = x(t) - c_1(t) = M_1(t) \\ r_2(t) = r_1(t) - h^2_{n_2} = r_1(t) - c_2(t) = M_2(t) \\ \vdots \\ r_n(t) = r_{n-1}(t) - h^2_{n_n} = r_{n-1}(t) - c_n(t) = M_n(t) \end{cases} \tag{4.13}$$

Solving for the nth residual yields

$$r_n(t) = x(t) - \sum_{k=1}^{n} h^k_{n_k}(t) = x(t) - \sum_{i=1}^{n} c_i(t)$$

as expected from (4.1).

The physical meaning of the decomposition becomes evident. As shown in (4.13), at each step of the process, the residue $r_j(t)$ becomes the *local* mean envelope of the previous step. Although these local mean signals carry only partial information, they can be used to describe overall system motion.

Motivated by these objectives, the following general nonstationary model has been proposed to describe the time evolution of the measured data $x(t)$ [5]:

$$\hat{x}(t) = M(t) + h_T(t) \tag{4.14}$$

where $h_T(t)$ is the fast total fluctuating component superimposed on a slow time-varying mean, and M is the slow, time-varying mean of the signal.

Substituting (4.8) into (4.11) and (4.12) and simplifying result in

$$\sum_{i=1}^{n} (M_{iT}(t)) = x(t) - c_1(t) + \sum_{i=2}^{n} (M_{i-1T}(t) - c_i(t))$$

or

$$\sum_{i=1}^{n} (M_{iT}(t)) = x(t) - \sum_{i=1}^{n} c_i(t)$$

It follows that the original signal, $x(t)$, can be approximated as

$$x(t) = M_T + \sum_{i=1}^{n} c_i(t)$$

where

$$M_T = \sum_{i=1}^{n} (M_{iT}(t)) \tag{4.15}$$

is the total mean.

Conventional HHT analysis has some limitations when used for monitoring of closely spaced modes and may result in unphysical modes.

4.3.2 Wavelet analysis

The details of the wavelet method have appeared elsewhere [19–21], and only a brief description will be given here.

Following Lijuan *et al.* [22], the discrete wavelet transform, W_ψ, of a signal $x(t)$ is given by

$$W_\psi(2^j, b) = \frac{1}{\sqrt{2^j}} \sum_{m=0}^{N-1} x(m)\psi^*\left(\frac{m-b}{2^j}\right)$$

where ψ is an appropriately chosen wavelet, the * indicates complex conjugate, $W_\psi(2^j,b)$ represents the wavelet coefficients, $\psi^*(m-b)/2^j$ is the conjugate of the mother wavelet with scaling or dilation factor 2^j, and b is the translation parameter.

The wavelet and scaling atoms of the discrete wavelet transform at resolution level j and location k are given by

$$\psi_{jk}^{i}(t) = 2^{-j/2}\psi^{i}(2^{-j}t - k)$$

and

$$\phi_{jk}^{i}(t) = 2^{j/2}\phi(2^{j}t - k)$$

Once the mother functions are known, a signal $x(t)$ can be represented as

$$x(t) = \sum_{k=-\infty}^{\infty} a_{o,k}\Phi_{o,k}(t) + \sum_{m=0}^{\infty}\sum_{k=-\infty}^{\infty} d_{m,k}\psi_{m,k}(t)$$

where

$$\begin{cases} a_{o,k} = \int_{-\infty}^{\infty} x(t)\phi_{o,k}(t)dt \\ d_{m,k} = \int_{-\infty}^{\infty} x(t)\psi_{o,k}(t)dt \end{cases}$$

The signal $x(t)$ can then be recovered by the inverse continuous wavelet transform as follows:

$$x(t) = W_{\psi}(2^{j}, b)\frac{1}{\sqrt{2^{j}}}\sum_{m=0}^{N-1} x(m)\psi^{*}\left(\frac{m-b}{2^{j}}\right) \tag{4.16}$$

The above analysis leads to a multiresolution decomposition of the signal, of the form

$$x(t) = \sum_{j=1}^{J} d_{j}(t) + d_{J}(t)$$

where $d_j(t)$ denotes the detail function at decomposition level j, J represents the total number of decomposition levels, and $d_J(t)$ represents the approximation function, which represents the trend of $x(t)$.

Application of wavelet analysis to power system measured data is described in recent works [14, 15].

4.3.2.1 Wavelet phase difference

Recently, the importance of phase analysis has been recognized by several researchers [21]. Following Marczak and Gómez [23], let the continuous wavelet transform of $x(t)$ be written in the form

$$W_{\psi}(2^{j}, b) = \frac{1}{\sqrt{|s|}}\int_{-\infty}^{\infty} x(t)\psi^{*}\left(\frac{t-\tau}{s}\right)d\tau \tag{4.17}$$

Let now x and y be two functions; the wavelet cross-spectrum is defined as

$$W_{xy,\psi}(s,\tau) = W_{x,\psi}(s,\tau)\, W^{*}_{y,\psi}(s,\tau)$$

or

$$\begin{aligned} W_{xy,\psi}(s,\tau) &= \mathrm{Re}\big(W_{xy,\psi}(s,\tau)\big) + \mathrm{Im}\Big(W^{*}_{xy,\psi}(s,\tau)\Big) \\ &= \big|W_{xy,\psi}(s,\tau)\big| \angle \phi_{xy,\psi}(s,\tau) \end{aligned}$$

where $\mathrm{Re}(W_{xy,\psi}(s,\tau))$ denotes the wavelet co-spectrum and $\mathrm{Im}(W_{xy,\psi}(s,\tau))$ is the wavelet quadrature spectrum. Avdaković *et al.* [21] define the phase angle of the spectrum as

$$\phi_{x,y,\psi}(\tau,s) = \arctan\left(\frac{\mathrm{Im}(W_{x,y,\psi},(\tau,s))}{\mathrm{Re}(W_{x,y,\psi},(\tau,s))}\right) \tag{4.18}$$

Similar to the phase angle in (4.12), the above expression can be used to extract phase relationships among modal components.

4.3.2.2 Relationship with EMD

It has been recently noted that synchrosqueezed wavelet transforms provide a similar decomposition to the EMD [16] resulting in a decomposition of a signal as a superposition of a finite number of approximately harmonic components.

It is observed that the EMD can be reinterpreted in terms of a nonlinear operator as in [24].

$$x(t) = \sum_{j=1}^{J_o+1} (P_R, x)(t) = x(t) \quad \text{for some } J_o \tag{4.19}$$

where

$$c_j(t) = (P_R, x)(t) \quad \text{and} \quad s_{J_o}(t) = (P_{RJ_o+1}, x)(t)$$

and

$$x(t) = \sum_{j=1}^{J_o} c_j(t) + s_{J_o}(t)$$

This provides a connection between wavelet and HHT analysis and paves the way for alternative techniques that exploit the benefits of both techniques.

Several combined approximations can be derived by combining wavelet analysis with Hilbert analysis. Table 4.3 summarizes some related approaches.

A useful overview of other linear and nonlinear analysis techniques applied to power system data is given in [2].

Table 4.3 Modal decomposition methods

Method	Reference
Wavelet decomposition + Hilbert analysis	Messina *et al.* [5]
L_p periodogram + Hilbert analysis	Lauria and Pisani [17]
EMD + Teager–Kaiser	Barocio *et al.* [25]
Filters + Teager–Kaiser	Kamwa *et al.* [26]
EMD + Wigner distribution	Palmer [27]

4.3.3 The Teager–Kaiser operator

The TKEO is a nonlinear operator developed to track the instantaneous energy content of speech signals using the concept of energy from a simple harmonic motion [28].

Consider a simple mass-spring system described by the mass normalized equation of motion

$$m\ddot{x} + c\dot{x} + kx + f = 0 \tag{4.20}$$

where x, $\dot{x}$, and $\ddot{x}$ denote displacement, velocity, and acceleration, respectively, k is a spring constant, m is mass, and c is the damping coefficient.

Assuming that nonlinear forces are relatively small, the solution of (4.20) is given by

$$\begin{aligned} x(t) &= a(t)\cos(\omega t + \theta) \\ \dot{x}(t) &= -a(t)\omega\sin(\omega t + \theta) + \cos(\omega t + \theta)\dot{a}(t) \end{aligned} \tag{4.21}$$

where $a(t) = A^{-\sigma t}$ is the time-varying amplitude, $\omega^2 = \omega_o^2 - \sigma^2 = g/m - b^2/4m^2$, $\omega_i^2 = \omega_o^2 - \sigma^2 = g/m - b^2/4m^2$ is the natural frequency of the oscillator, and $\sigma = b/2m$ is the energy dissipation rate.

The mass normalized instantaneous energy in the system is defined by

$$E_T(t) = \frac{1}{2}\omega^2 x^2(t) + \frac{1}{2}\dot{x}^2(t) \tag{4.22}$$

Substitution of (4.3) into (4.4) and subsequent simplification yield

$$E_T(t) \approx \omega^2 (Ae^{-\sigma t})^2 = (\dot{x}(t))^2 - x(t)\ddot{x}(t) \tag{4.23}$$

For a continuous signal $x(t)$, the Teager–Kaiser Energy Operator (TKEO), $\Psi(x(t))$, is defined as [28]

$$\Psi(x(t)) \approx (\dot{x}(t))^2 - x(t)\ddot{x}(t) \tag{4.24}$$

where the operator $\Psi(x(t))$ has the units of energy.

Application of this criteria to (4.32) yields

$$\Psi(x(t)) \approx a^2\omega^2$$

and

$$\Psi(\dot{x}(t)) \approx a^2\omega^4$$

Combining these equations and solving for the amplitude and frequency result in

$$\begin{cases} A = \dfrac{\Psi(x(t))}{\sqrt{\Psi(\dot{x}(t))}} \\ \omega = \sqrt{\dfrac{\Psi(\dot{x}(t))}{\Psi(x(t))}} \end{cases} \tag{4.25}$$

The discrete form of the TKEO is

$$\begin{cases} |A| = \dfrac{2\Psi(x(k))}{\sqrt{\Psi(x(k+1) - x(k-1))}} \\ \omega = \arcsin\sqrt{\dfrac{\Psi(x(k+1) - x(k-1))}{4\Psi(x(k))}} \end{cases} \tag{4.26}$$

where k is the discrete time. This equation is referred to as the discrete-time energy separation algorithm (DESA).

Several practical criteria for computing the TKEO have been discussed in the literature [25]. Traditional DESA algorithms work well for noiseless signals, but have difficulties incorporating noise and require that a signal is decomposed into mono-component signals.

The use of filters to avoid the EMD has been previously discussed by Kamwa *et al.* [26].

4.3.4 Dynamic harmonic regression

Dynamic harmonic regression (DHR) has recently emerged as a method for dealing with nonlinear processes. These models are of the general form [29, 30]

$$x(t) = T(t) + S(t) + C(t) + e(t), \quad e(t) = N(0, \sigma_e^2) \tag{4.27}$$

where $x(t)$ is the observed time series, t denotes time, and $T(t)$, $S(t)$, and $C(t)$ represent the trend, quasi-cyclical, and stochastic components, respectively; $e(t)$ is an irregular component normally distributed Gaussian sequence with zero mean value and variance σ_e^2.

In its simplest form, the DHR model can be written as

$$x(t) = \sum_{j=0}^{R} s_j(t) + e(t) = \sum_{j=0}^{R} \underbrace{\left[a_j(t)\cos(\omega_j t) + b_j(t)\sin(\omega_j t)\right]}_{s_j(t)} + e(t) \tag{4.28}$$

where $a_j(t)$ and $b_j(t)$ are assumed to be stochastic time-varying parameters that follow a generalized random walk process, and $e(t)$ is a residual series assumed to represent noise in the time series. The parameter R determines the number of harmonic regressions that are allowed in the model (4.28).

In this representation, each $a_j(t)$, $b_j(t)$ is a stochastic time-variable parameter (TVP), and the ω_j, $j = 1, \ldots, R$, are the fundamental and harmonic frequencies associated with the periodicity in the time series; a_o is a slowly varying parameter or a trend obtained assuming $T(t) = S(t \to \infty) = a_o(t)\cos(0 \cdot t) + b_o(t)\sin(0 \cdot t)$ in (4.19). As a result, nonstationarity is allowed in the various components.

To introduce the model, note that the measured components y_t in (4.27) can be expressed in the general form

$$y_{j_t} = \begin{bmatrix} \cos(\omega t) & \sin(\omega t) \end{bmatrix} \begin{bmatrix} a_{j_t} \\ b_{j_t} \end{bmatrix} + e_{j_t}$$

It follows that if the frequencies ω_j can be estimated, the time-varying parameters, a_j, b_j, can be determined in a straightforward manner. The instantaneous amplitude and phase of each TVP can be computed from

$$\begin{cases} A_j(t) = \sqrt{a_j^2(t) + b_j^2(t)} \\ \varphi_j(t) = \arctan \dfrac{b_j(t)}{a_j(t)} \end{cases}, \quad j = 1, \ldots, R$$

In the standard DHR model, the frequency values ω_j, $j = 1, \ldots, R$ are obtained from the autoregressive (AR) spectrum, the Fourier transform, or wavelet analysis, but other methods can be used including HHT or Bayesian analysis. In the latter case, the frequencies can be included in the model as unknown parameters.

The DHR model estimates all the parameters in (4.28) simultaneously using a combination of forward-recursive filtering followed by backwards recursive fixed interval smoothing. Zavala and Messina [30] outlined two proposals for estimating the frequency components within the framework of DHR analysis.

More general representations include generalized additive models. Estimates of the states or time-varying parameters are then obtained using an optimal estimation method based on the Kalman filter.

4.3.4.1 State space modeling framework

Time varying models of the form (4.28) assume that the slope and variance of the time series change over time. In the analysis that follows, a random walk plus noise model is adopted in which the evolution of each of the $2R + 1$ parameters is characterized by the following two variables:

1. The amplitude l_{j_t}, and
2. The slope or drift, d_{j_t}

Under these assumptions, let the stochastic state vector be defined as $x_{j_t} = [\, l_{j_t} \quad d_{j_t} \,]^T$. The state space representation of the model (4.27), (4.28) can be written as a Gaussian state space model of the form [30]

$$\mathbf{x}_j(t) = \mathbf{F}_j\mathbf{x}_j(t-1) + \mathbf{G}_j\boldsymbol{\eta}_j(t) \tag{4.29}$$

where $\boldsymbol{\eta}_j(t) = \left[\, v_j(t) \quad \xi_j(t) \,\right]^T$, $v_j(t) \sim w.n.N(0, \sigma^2_{v_j})$, $\xi_j(t) \sim w.n.N(0, \sigma^2_{\xi_j})$, and

$$\mathbf{F}_j = \begin{bmatrix} \alpha_j & \beta_j \\ 0 & \gamma_j \end{bmatrix}; \quad \mathbf{G}_j = \begin{bmatrix} \delta_j & 0 \\ 0 & 1 \end{bmatrix}$$

Once the dynamic model has been expressed in state-space form, the Kalman filter can be used to estimate the state or time-varying parameters as discussed below.

4.3.4.2 Kalman filter and smoothing algorithms

To introduce the adopted model, consider the problem of estimating a state vector $\mathbf{x}(k)$ associated with a stochastic dynamic system modeled by the simple Gauss–Markov process:

$$\text{State equations:} \quad \mathbf{x}_j(t) = \mathbf{F}_j\mathbf{x}_j(t-1) + \mathbf{G}_j\boldsymbol{\eta}_j(t) \tag{4.30a}$$

$$\text{Observation equations:} \quad \mathbf{y}_j(t) = \mathbf{H}_j\mathbf{x}_j + \boldsymbol{\xi}(t) \tag{4.30b}$$

where $\mathbf{y}_j$ is a $px1$ dimensional vector of observations that are linearly related to the state vector $\mathbf{x}_j$ by the matrix $\mathbf{H}_j$, and $\boldsymbol{\eta}_j, \boldsymbol{\xi}_j(t)$ are zero-mean statistically independent white-noise disturbance vectors with possibly time-variable covariance matrices $\mathbf{Q}_j$ and $\mathbf{R}_j$, respectively.

Having expressed the dynamic model in state-space form, the Kalman filter can be used to estimate the state $\mathbf{x}$. More formally, given a set of measurements, $x(t_1)$, $x(t_2), \ldots,$ $x(t_N)$, the optimal estimate $\hat{x}(t+1)$ of $\hat{x}(t)$ can be obtained by minimizing the expected value of the magnitude of the error

$$\min_{\hat{\mathbf{x}}}\left\{E\left[\left|\mathbf{x}_t - \mathbf{x}_{t|\tau}\right|^2\right]\right\}$$

where the subscript $t|\tau$ refers to an estimate at time t given information up to and including time τ.

4.3.4.3 Estimation of the time-variable parameters

DHR estimates the time-varying parameters using a two-step (prediction-correction) Kalman filter followed by a fixed-interval smoothing algorithm. The process for optimal state estimation can be described by the following equations [29, 31]:

(a) Prediction

$$\begin{aligned} \hat{\mathbf{x}}_{t|t-1} &= \mathbf{F}\hat{\mathbf{x}}_{t-1|t-1} \\ \mathbf{P}_{t|t-1} &= \mathbf{F}\mathbf{P}_{t-1|t-1}\mathbf{F}^T + \mathbf{G}\mathbf{Q}\mathbf{G}^T \end{aligned} \tag{4.31}$$

(b) Correction

$$\begin{aligned}
\Lambda_t &= \mathbf{x}_t - \mathbf{H}\hat{\mathbf{x}}_{t|t-1} \\
\mathbf{S}_t &= \mathbf{H}\mathbf{P}_{t|t-1}\mathbf{H}^T + \mathbf{R}_t \\
\mathbf{K}_t &= \mathbf{P}_{t|t-1}\mathbf{H}^T\mathbf{S}_t^{-1} \\
\hat{\mathbf{x}}_{t|t} &= \hat{\mathbf{x}}_{t|t-1} + \mathbf{K}_t\Lambda_t \\
\mathbf{P}_{t|t} &= (\mathbf{I} - \mathbf{K}_t\mathbf{H})\mathbf{P}_{t|t-1}
\end{aligned} \tag{4.32}$$

where $\mathbf{Q} = diag(\begin{bmatrix}\sigma_v^2 & \sigma_\xi^2\end{bmatrix})$, with initial conditions $\hat{\mathbf{x}}_o$ and $\mathbf{P}_o$, and the notation $\hat{\mathbf{x}}_{t|t-1}$ is used to indicate the estimate of $\mathbf{x}(t)$ given the observations $x(0), x(1), \ldots, x(t-1)$.

After the filtering stage, a fixed interval smoother is used to update (correct) the filter estimated state $\hat{\mathrm{x}}_{t|t}$. In this case, using the output of the Kalman filter, smoothing takes the form of a backward recursion for $t = N, \ldots, 1$, operating from the end of the sample set to the beginning:

$$\begin{aligned}
\hat{\mathbf{x}}_{t+1|N} &= \hat{\mathbf{x}}_{t+1|t+1} - \mathbf{P}^*_{t+1|N}\mathbf{F}^T_{t+1}\boldsymbol{\lambda}_{t+1} \\
\hat{\mathbf{x}}_{t+1|N} &= \hat{\mathbf{x}}_{t+1|t+1} - \mathbf{P}^*_{t+1}\mathbf{F}^T_{t+1}\boldsymbol{\lambda}_{t+1} \\
\mathbf{P}^*_{t|N} &= \mathbf{P}^*_{t|t} + \mathbf{P}^*_t\mathbf{F}\mathbf{P}^{*\ -1}_{t+1|t}\left[\mathbf{P}^*_{t+1|N} - \mathbf{P}^*_{t+1|t}\right]\mathbf{P}^{*\ -1}_{t+1|t}\mathbf{F}\mathbf{P}^*_{t|t} \\
\boldsymbol{\lambda}_t &= \left(\mathbf{I} - \mathbf{H}^T_{t+1}\mathbf{R}^{-1}_{t+1}\mathbf{H}_{t+1}\mathbf{P}_{t+1|t+1}\right)\mathbf{F}^T_{t+|}\boldsymbol{\lambda}_{t+1} \\
&\quad - \mathbf{H}^T_{t+1}\mathbf{R}^{-1}_{t+1}\left(\mathbf{y}_{t+1} - \mathbf{H}_{t+1}\hat{\mathbf{X}}_{t+1|t+1}\right)
\end{aligned}$$

Observe that the algorithm requires specifying the initial condition x_o and its associated error covariance P_o.

4.3.4.4 Trend extraction

Compared to more traditional approaches, unobserved component time-series models have the potential to include frequency information, local trends, and oscillatory and irregular components.

In practical applications the local linear trend model can be placed into the Gaussian state-space form

$$\begin{aligned}
\mu_t &= \mu_{t-1} + \beta_t + \xi_t \\
\beta_t &= \beta_{t-1} + \zeta_t
\end{aligned} \tag{4.33}$$

where μ_t is a linear trend or level, ξ denotes a white noise process associated with the trend, β is the slope of the trend, and ζ is a white-noise process associated with the slope of the trend; the subscript t denotes the information available at time t. The local-linear trend model reduces to a random walk model when $\beta = 0$. Note that this model admits a state representation of the form (4.29). A similar procedure can be used to represent periodic and other components.

In [29] a procedure to estimate the signal-to-noise ratio (SNR) in the frequency domain has been suggested. The method is based on the fact that the pseudo-spectrum of the DHR model of the $R+1$ frequency components in the model can be estimated as

$$f_y(\omega, \boldsymbol{\sigma}^2) = \frac{1}{8\pi}\sum_{j=0}^{R}\left[\frac{\sigma^2_{\omega_j}}{\left(1-\cos(\omega+\omega_j)\right)^2}\frac{\sigma^2_{\omega_j}}{\left(1-\cos(\omega-\omega_j)\right)^2}\right] + \frac{\sigma^2}{2\pi} \tag{4.34}$$

where $\boldsymbol{\sigma}^2 = \begin{bmatrix}\sigma^2 & \sigma^2_{\omega_o} & \sigma^2_{\omega_1} & \cdots & \sigma^2_{\omega_R}\end{bmatrix}$. Here, $\sigma^2_{\omega_o}$ is the variance associated with the zero frequency term (the trend), the $\sigma^2_{\omega_i}$, $i = 1, \ldots, R$ are the variances associated with the harmonic components, and σ is the variance of $e(t)$.

The SNR in $\mathbf{Q}(t)$ is defined as

$$\text{SNR} = \frac{\sigma^2_\xi}{\sigma^2_\varepsilon}$$

An estimate of the initial trend and the SNR matrix are needed to initiate the algorithm, and the SNR must be estimated separately.

Under the assumption that each element $a_j(t)$, $b_j(t)$ follows an autoregressive (AR) model, the time-varying parameters can be estimated by minimizing the functional

$$J = \sum_{j=0}^{R}\left[f_y(\omega_j) - \hat{f}_y(\omega_j, \boldsymbol{\sigma}^2)\right]^2 \tag{4.35}$$

where $f_y(\omega_j)$ is the spectrum of $y(t)$.

Central to this procedure is the estimation of the dominant harmonic frequencies ω_j, $j = 0, \ldots, R$. Then, the unknown time-varying parameters in (4.28) can be estimated from least-squares optimization of the expected value of the magnitude of the error.

Alternatives to (4.34) are the use of near real-time nonlinear and nonstationary methods such as HHT or similar techniques.

Estimation of the variances is then straightforward:

Dynamic harmonic regression procedure

Given a set of simultaneously recorded signals $x_k(t)$, $k = 1, \ldots, m$:

1. Estimate the dominant harmonic frequencies $\omega_j, \forall j = 1, 2, \ldots, R$ from the AR spectrum.
2. Compute the unknown vector of parameters $\boldsymbol{\sigma}^2$ from least-squares optimization of the functional

$$\min_{\hat{x}}\left\{E\left[\left|\mathbf{x}_t - \mathbf{x}_{t|T}\right|^2\right]\right\}$$

3. Once these parameters are estimated, the time-varying trends and harmonic components can be obtained using the Kalman filter using (4.31) and (4.32).
4. Use the optimal fixed interval smoothing to determine optimal estimates for the time-variable parameters.
5. Reconstruct the measured signal from the selected parameters as

$$\hat{x}(t) = \sum_{j=0}^{R}\left[a_j(t)\cos(\omega_j t) + b_j(t)\sin(\omega_j t)\right] + e(t)$$

Compared to other approaches, time-series models have the potential to include frequency information, local trends, and oscillatory and irregular components. Approaches are needed to generalize the system model to include damping information.

4.3.4.5 Forecasting

The state space estimation based on optimal Kalman filter together with fixed interval smoothing is well suited for handling missing observations, forecasting, and outliers. Reference 30 describes the application of DHR to predict behavior from measured power system data.

The following example compares the application of HHT analysis and DHR to extract specific system behavior.

Example 4.1 Simultaneous trend extraction To motivate the application of the technique consider the simulated bus voltage magnitudes for the Westinghouse test system in Section 3.6. In the developed algorithms, time-varying trends are estimated simultaneously for the simulated bus voltage signals.

Figure 4.3a shows the time evolution of selected voltage measurements, along with the instantaneous mean extracted using the above procedure. Figure 4.3b shows the corresponding slope of the trend β.

Of note, the analysis suggests that the slope of the trend can be used to identify coherent behavior in measured data.

Example 4.2 Application to measured data As a second example of the application of the above ideas, consider the recorded tie-line power signal in Figure 4.4 based on the phasor measurement unit (PMU) data. The signal exhibits a slow trend and switching event at about 175 s into the measurement.

Using the EMD method, the local means were computed based on the procedure developed above. At each iteration, the local means were computed using

$$M_{iT}(t) = \sum_{k=1}^{n} \left(m_k^j(t)\right)$$

Hilbert analysis results in eight IMFs and a slow trend or residual. Figure 4.5a shows the three first IMFs and the residual, Figure 4.5b shows the extracted local means, and Figure 4.5c depicts the mean squared error $MSE_i = |M_i(t) - M_{i-1}(t)|$ as a function of the iteration level, i. Typically, as shown in Figure 4.5, the MSE decreases to zero in a few iterations (three to five iterations for most signals).

Also of interest, Figure 4.6a compares the mean estimate obtained from DHR analysis with the mean from HHT. For the purposes of rigorous comparison, results are found to correlate very well. Figure 4.6b shows the detrended signal. Using this approach, localized events in time are singled out.

4.4 Mutivariate multiscale analysis

The above analysis extends readily to the mustiscale case. Attention is now turned to two well-developed methods: multisignal Prony analysis and the Koopman mode decomposition.

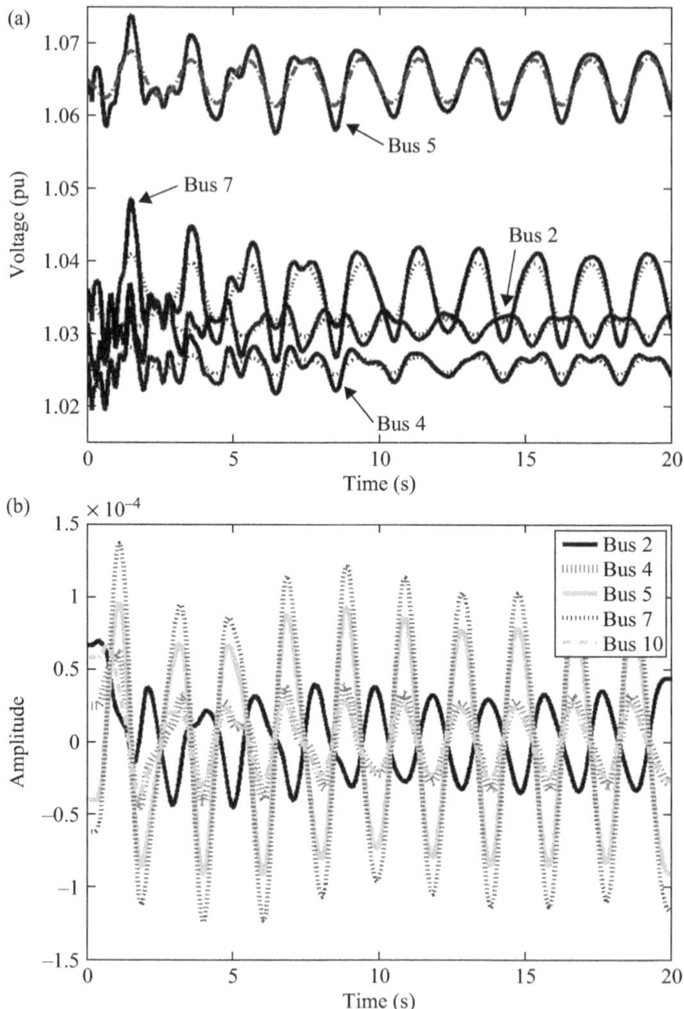

Figure 4.3 Slope and trend derivative as a function of time (a) simulated signals and associated trends, μ_t (dotted lines); trend derivatives, $d\beta/dt$

4.4.1 Multi-signal Prony analysis

Multisignal Prony analysis is the prevailing method for modal analysis of various simultaneous measurements [2]. Prony analysis, however, may not be efficient at describing large-scale system dynamics and is subjected to some of the same limitations of the univariate methods.

A detailed description and derivation of Prony models are given by Sanchez-Gasca *et al.* [2] and Trudnowski *et al.* [32] for a detailed description of this method.

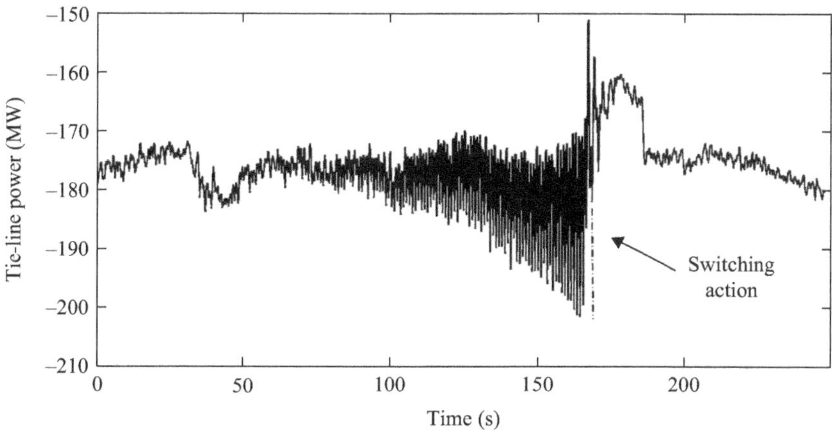

Figure 4.4 Measured signal used in the example

4.4.2 Koopman analysis

Another class of global monitoring systems is the Koopman operational mode introduced in [33, 34].

The Koopman method has its roots in the field if dynamic systems. The method assumes that the dynamic behavior of a nonlinear process can be written as a discrete time system or dynamic map. Following Susuki and Mezic [33], consider a discrete-time system evolving on an N-dimensional manifold M:

$$\mathbf{x}_{k+1} = \mathbf{f}(\mathbf{x}_k) \tag{4.36}$$

with $k = 0, 1, 2, \ldots, N$ where k is an integer index, and $\mathbf{x}$ is the N-dimensional vector of states.

Let now $g(\boldsymbol{x})$: $M \to \mathbb{R}$ be any scalar-valued function (a measurement of the state or observable) of dimension $p < N$. The Koopman operator, U, is a linear operator that maps g into a new function

$$Ug(\mathbf{x}) = g(\mathbf{f}(\mathbf{x})) \tag{4.37}$$

The key idea behind Koopman analysis is to study the system dynamics (4.36) from measured data using the eigenspectrum of U. Assume to this end that φ_j and λ_j denote the eigenfunctions and eigenvalues (Koopman modes) of the Koopman operator, respectively, given by

$$U\varphi_j(\mathbf{x}) = \lambda_j \varphi_j(\mathbf{x}), \quad j = 1, 2, \ldots \tag{4.38}$$

where for N sufficiently long, the Koopman eigenfunctions form an orthonormal expansion basis [34, 35].

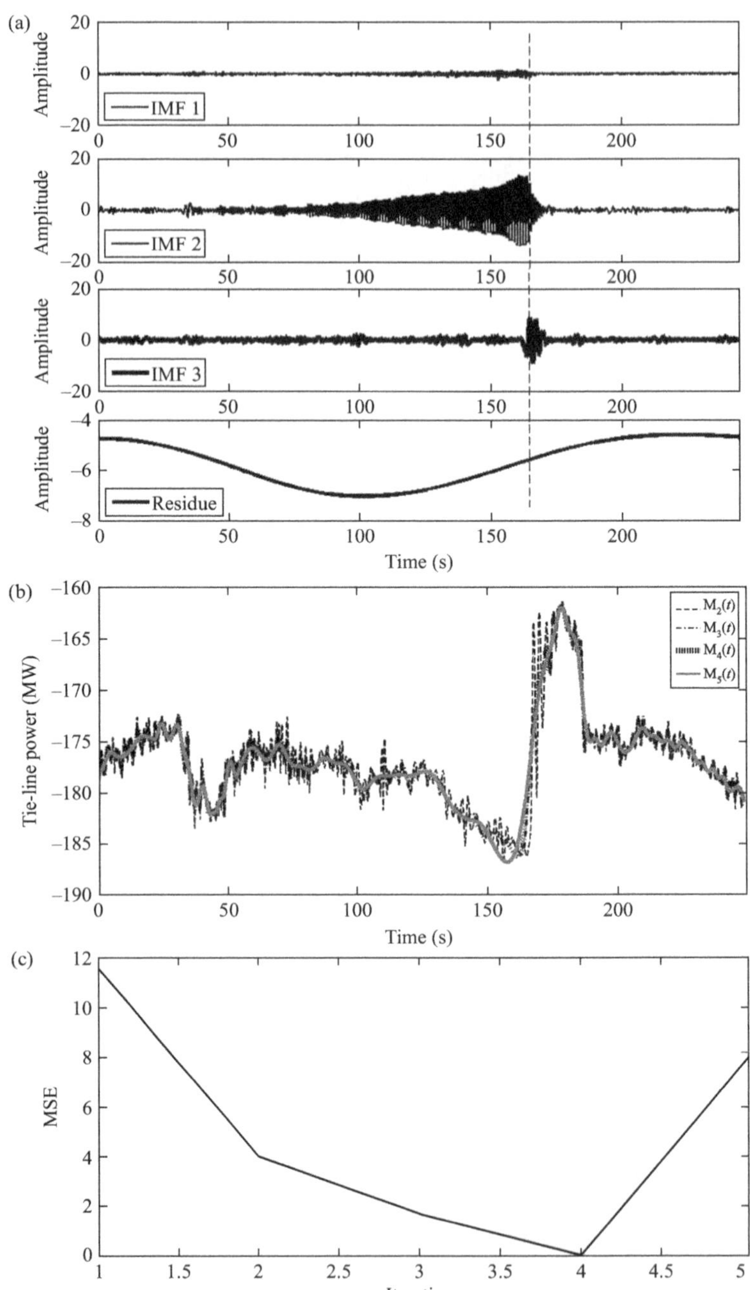

Figure 4.5 Minimum squared error, MSE, as a function of the EMD decomposition level: (a) extracted IMFs; (b) extracted local means; (c) mean squared error

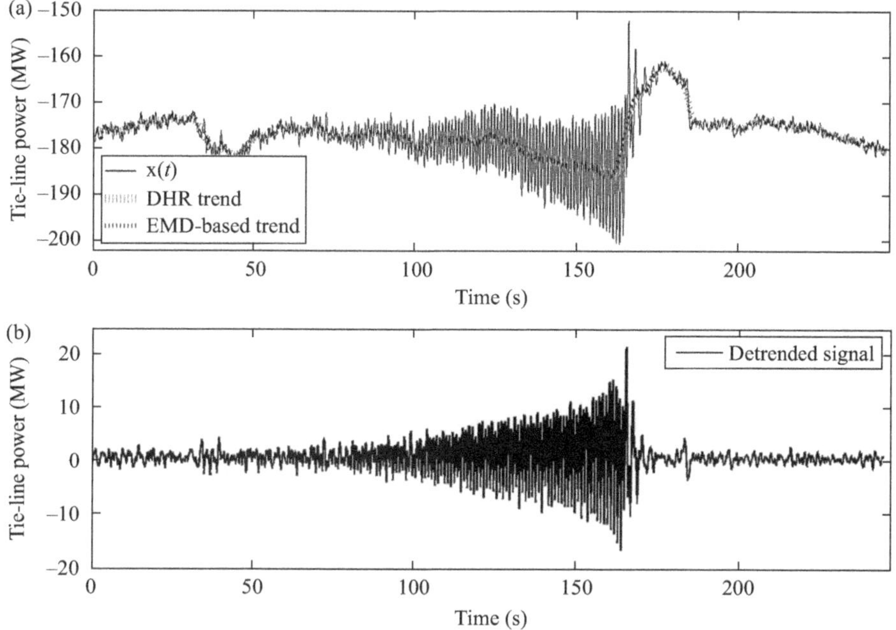

Figure 4.6 Comparison of detrending technique

In practical applications, one is interested in functions $g(\mathbf{x}) = [g_1(\mathbf{x})\ g_2(\mathbf{x})\dots g_p(\mathbf{x})] = M \rightarrow p$, with $p < N$. Assuming further that each of the components of g lie within the span of the eigenfunctions φ_j, the time evolution of the functions $g_1(\mathbf{x}_k)$ can be expanded as

$$\boldsymbol{g}(\boldsymbol{x}) = \sum_{j=1}^{\infty} \varphi_j(\boldsymbol{x})\boldsymbol{v}_j \tag{4.39}$$

and

$$\boldsymbol{x}_k = \boldsymbol{g}(\boldsymbol{x}_k) = \sum_{J=1}^{\infty} U^k \varphi_j(\boldsymbol{x}_0)\boldsymbol{v}_j = \sum_{J=1}^{\infty} \lambda_j^k \varphi_j(\boldsymbol{x}_0)\boldsymbol{v}_j \tag{4.40}$$

where use has been made of (4.37).

Physically, (4.40) indicates that the observable $\boldsymbol{g}(\boldsymbol{x}_k)$ is decomposed into vector coefficients, $\boldsymbol{v}_j$, called Koopman modes whose temporal behavior is given by the associated eigenvalues λ_j; the phase of the eigenvalues determines its frequency, while its modulus determines the growth rate. The magnitude $\varphi_j(\boldsymbol{x}_0)\boldsymbol{v}_j$ is used as a measure of the relative participation of a mode to the modal decomposition [35].

Analytical approaches to compute Koopman modes based on Arnoldi-like algorithms have been developed and tested on measured data of the form (4.36).

Following the same notation as used by Susuki and Mezic [33], consider the finite-time m-by-$N-1$ data (observation) matrix

$$\mathbf{P} = \hat{\mathbf{X}} = [\mathbf{P}_o \quad \mathbf{P}_1 \quad \cdots \quad \mathbf{P}_{N-1}] = \begin{bmatrix} p_{10} & p_{11} & \cdots & p_{1,N-1} \\ p_{20} & p_{21} & \cdots & p_{2,N-1} \\ \vdots & \vdots & \ddots & \vdots \\ p_{m0} & p_{m1} & \cdots & p_{m,N-1} \end{bmatrix}$$

where m is the number of sensors or PMUs, $g(\boldsymbol{x}_0) = \boldsymbol{P}_0$, and each data column, $\boldsymbol{P}_i$, has a similar interpretation to that in the observation matrix $\mathbf{X}$.

The computation of the Koopman modes can be summarized as follows [23]:

Pseudo algorithm for Koopman decomposition of an ensemble of observations

1. Find constants c_j such that

$$\boldsymbol{r} = \boldsymbol{P}_{N-1} - \sum_{j=0}^{N-2} c_j \boldsymbol{P}_j$$

$$\boldsymbol{r} \perp \{\boldsymbol{P}_0 \quad \boldsymbol{P}_1 \quad \ldots \quad \boldsymbol{P}_{N-1}\}$$

2. Determine the eigenvalues (Ritz values) $\lambda_1, \lambda_1, \ldots, \lambda_{N-1}$ of the companion matrix $\mathbf{C}$:

$$\mathbf{C} = \begin{bmatrix} 0 & & & & c_o \\ 1 & 0 & & & c_1 \\ & \ddots & \ddots & & \vdots \\ & & 1 & 0 & c_{N-3} \\ & & 0 & 1 & c_{N-2} \end{bmatrix} = \mathbf{T}^{-1}\mathbf{\Lambda}\mathbf{T}$$

3. Define the Vandermonde matrix:

$$\mathbf{C} = \begin{bmatrix} 1 & \lambda_1 & \lambda_1^2 & \cdots & \lambda_1^{N-2} \\ 1 & \lambda_2 & \lambda_2^2 & \cdots & \lambda_2^{N-2} \\ 1 & \lambda_3 & \lambda_3^2 & \cdots & \lambda_3^{N-2} \\ \vdots & \vdots & & \ddots & \vdots \\ 1 & \lambda_{N-1} & \lambda_{N-1}^2 & \cdots & \lambda_{N-1}^{N-2} \end{bmatrix} \in \Re^{N-1xN-1}$$

4. Compute the Ritz vectors $\boldsymbol{v}_j$ in (4.39) as the columns of $\boldsymbol{V} = \boldsymbol{PT}^{-1}$. The Ritz vectors $\boldsymbol{v}_j$ approximate the terms $\varphi_j(x_o)\boldsymbol{v}_j$ in (4.39).

In this procedure, the constants c_j are determined solving the least-squares problem

$$\mathbf{0} = \mathbf{b} - \mathbf{Ac}$$

where $\boldsymbol{c} = [c_o \quad c_1 \quad c_2 \quad \ldots \quad c_{N-2}]^T$, with $\boldsymbol{b} \in \boldsymbol{R}^{N-1}$, with $b_{ij} = \boldsymbol{P}_i^T \boldsymbol{P}_{N-1}$ and $\boldsymbol{A} = \{A_{ij}\} \in \boldsymbol{R}^{(N-1)x(N-1)}$, with $A_{ij} = \boldsymbol{P}_i^T \boldsymbol{P}_{N-1}$.

The Koopman eigenfunctions are then obtained from matrix **T**.

Several observations are of interest here:

(a) The size of the eigenvalue problem may be very large $(N-1)$.
(b) Modal estimates are obtained for a given observation window. As a result, local changes in system behavior cannot be singled out.
(c) Mode shape estimates are complex-valued.
(d) Koopman analysis is sensitive to noise and other data characteristics.

Reference is made to Susuki and Mezic [33, 34] for recent discussions of Koopman analysis techniques including the various relations stated below. Variations to this method, which reduce the data matrix to the number of sensors, have been recently developed based on a dynamic decomposition algorithm [35]. Other recent approaches to global characterization of system behavior include multichannel ARMAX algorithms [36].

Compared with multisignal Prony analysis, the advantage of Koopman mode analysis is that nonlinearities can be taken into account without any approximation. Koopman methods, however, can be memory intensive (storage increases linearly with the number of samples) and expensive in terms of CPU time.

4.5 Response under ambient stimulus

Many studies suggest that random load variations can act to stochastically force the power system and excite the system electromechanical modes [37, 38].

In this section, a rigorous procedure to assess the impact of random system variation on system behavior is presented and outlined. This technique can be used to determine baseline information for model validation and power system health analysis.

4.5.1 Formulation of the model

The effects of random forcing on system behavior can be estimated using a linearized power system representation. Assume that, under small perturbations, the dynamic system of interest is fully described by the linear time-invariant model [39]:

$$\begin{aligned} \mathbf{x} &= \mathbf{Ax}(t) + \mathbf{Bu}(t) + \mathbf{F}\boldsymbol{\xi}(t), \quad \mathbf{x}(0) = \mathbf{x}_o \\ \mathbf{y} &= \mathbf{Cx}(t) + \mathbf{r}(t) \end{aligned} \tag{4.41}$$

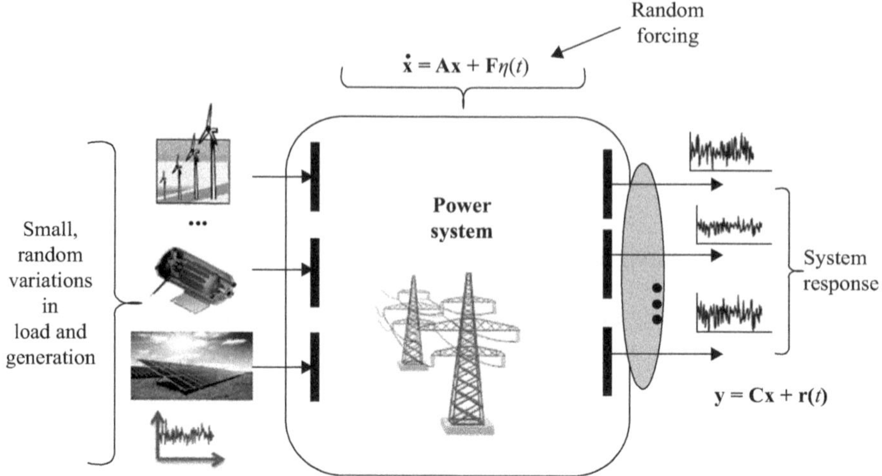

Figure 4.7 Conceptual power system representation for the analysis of stochastic forcing

where $\mathbf{x}(t)$ is an n-dimensional vector representing the state of the system, $\boldsymbol{\xi}$ is an m-dimensional vector representing the stochastic forcing, $\mathbf{y}(t)$ is the r-dimensional vector of outputs (the vector of measurements), and $\mathbf{r}(t)$ represents measurement noise. Matrix $\mathbf{A}$ represents the deterministic dynamics and matrix $\mathbf{F}$ represents the spatial distribution of the forcing; $\mathbf{A}$, $\mathbf{B}$ and $\mathbf{C}$ are matrices of appropriate dimensions.

Figure 4.7 gives a conceptual representation of the adopted model. Here, the inputs to the model are random load variations.

For simplicity, it is assumed that the state matrix $\mathbf{A}$ is time-independent and asymptotically stable, that is, all the eigenvalues of the state matrix have negative real parts.

Physically, the vector $\boldsymbol{\xi}$ represents random load and generation variations associated with renewable generation, and is assumed to be represented by a sequence of uncorrelated samples (white noise) with zero mean and unit variance, that is

$$\langle \vec{\boldsymbol{\xi}}(t) \rangle = 0$$

$$\langle \vec{\boldsymbol{\xi}}_i(t_m)\boldsymbol{\xi}_j(t_n) \rangle = R_{ij}\delta_{mn}$$

where angular brackets denote an ensemble average, δ_{mn} is the delta Kronecker function, and the R_{ij} are elements of the spatial covariance matrix of the noise.

In the section that follows, two analytical procedures to analyze the stochastic system performance are outlined. The first is based on the modal response to random load variations. The second approach is of interest to the analysis of ensembles of realizations using multivariate stochastic methods.

4.5.2 Modal response

Much insight into the nature of stochastic behavior can be obtained from linear analysis of the stochastic linear model. The solution of (4.41) with initial conditions $\mathbf{x}_o$ is given by

$$\mathbf{x}(t_o+\tau) = \underbrace{e^{\mathbf{A}\tau}\mathbf{x}(t_o)}_{\text{Deterministic response}} + \underbrace{\int_0^{\tau} e^{\mathbf{A}(t-\tau)}\xi(t+t_o)d\tau}_{\text{Random noise forcing}} \tag{4.42}$$

with the initial conditions $\mathbf{x}(t_o) = \mathbf{x}_o$, where τ is the lead time.

The first term on the RHS represents the effect of initial conditions and vanishes for a stable system; the second term represents the influence of random noise forcing on system behavior. When all the eigenvalues of the deterministic dynamics in $\mathbf{A}$ are negative, the first term tends to zero as $t \to \infty$, and the system response is given by the second term in (4.42). Moreover, the random response is linearly dependent on $\boldsymbol{\xi}$ and consequently is also Gaussian distributed.

To examine the stochastic growth of perturbations, let the forced solution of (4.42) be expressed as in [39, 40]:

$$\mathbf{x}(t_o+\tau) = \int_0^{\tau} e^{\mathbf{A}(t-\tau)}\mathbf{F}\boldsymbol{\eta}(t)d\tau \tag{4.43}$$

It follows that the variance associated with the stochastic forcing is given by

$$\left\langle \|\mathbf{y}(t)\|^2 \right\rangle = \left\langle \mathbf{C}\mathbf{F}^T \int_0^t e^{A(t-\tau)} e^{A^T(t-\tau')} d\tau\, \mathbf{C}^T \right\rangle \mathbf{F} = \mathbf{C}\mathbf{Q}(t)\mathbf{C}^T \tag{4.44}$$

where $\mathbf{Q}(t) = \int_0^t e^{A(t-\tau)}\mathbf{F}\mathbf{F}^T e^{A^T(t-\tau')}d\tau$.

Several conclusions can be drawn from this analysis:

1. The hermitian operator $\mathbf{Q}(t)$ accumulates the perturbation growth when all loads are stochastically excited.
2. For $\mathbf{A}$ constant, it follows that the system will reach a statistical steady state in which $\mathbf{Q}(\infty) = \lim_{t\to\infty}\mathbf{Q}(t)$ is a solution of a Lyapunov equation:

$$\mathbf{A}\mathbf{Q}_\infty + \mathbf{Q}_\infty\mathbf{A}^T = -\mathbf{F}\mathbf{F}^T \tag{4.45}$$

From the previous discussion, it also follows that the mean energy of the stochastic process can be extracted from the covariance matrix. Its expected value is given by

$$E(t) = \left\langle \mathbf{y}^T(t)\mathbf{y}(t) \right\rangle = \text{trace}\left[\mathbf{C}^T\mathbf{Q}(t)\mathbf{C}\right]$$

The interested reader is referred to research by Fontane *et al.* [40] and Farell and Ioannou [41] for a detailed derivation.

4.5.3 Ensemble system response

Direct application of the above analysis to complex system representations, however, has some limitations:

- Noise may induce nonlinearities.
- The accuracy of modal estimates is dependent upon the number of samples.
- The spatial patterns φ are stationary.
- The statistical basis $\mathbf{\Phi}$ only contains information that is present in the snapshots.

Let the solution of the system (4.42) be expressed as [42]

$$\mathbf{x}(t+\Delta t) = \underbrace{[1+\mathbf{B}\Delta t]\mathbf{x}(t)}_{\text{Deterministic part}} + \underbrace{\left(\sigma\sqrt{\Delta t}\right) randn}_{\text{Stochastic part}} \tag{4.46}$$

where Δt is the integration step, and the other variables have the usual interpretation.

These solutions are used as a reference for long-term electromechanical simulation on ambient noise effects on a power system [38]. When combined with a multivariate statistical analysis technique, the ensemble of realizations can be used to extract modal information under more general operating conditions.

The procedure can be summarized as follows:

Modal analysis of the state representation

Given a small signal model of the form (4.41):

1. Model load variations at selected system locations.
2. Generate an ensemble of system realization of the undamaged system using (4.43) or (4.46).
3. Compute stochastic modes using the statistical approaches in Chapter 3.
4. Obtain the dominant (optimal) forcing as well as zones of modal activity, levels of healthy system behavior, etc.

Since the spatial structure of the stochastic forcing plays an important role in exciting system variability on different scales, critical loads and transmission paths associated with major inter-area modes can be determined.

4.6 Application to measured data

The above procedures are illustrated on measured frequency data from a large power system. Figure 4.8 shows a schematic of the system showing the location of measurement points together with the measured data. The system is composed of six regional systems; for purposes of evaluation, a frequency measurement is taken from each regional system.

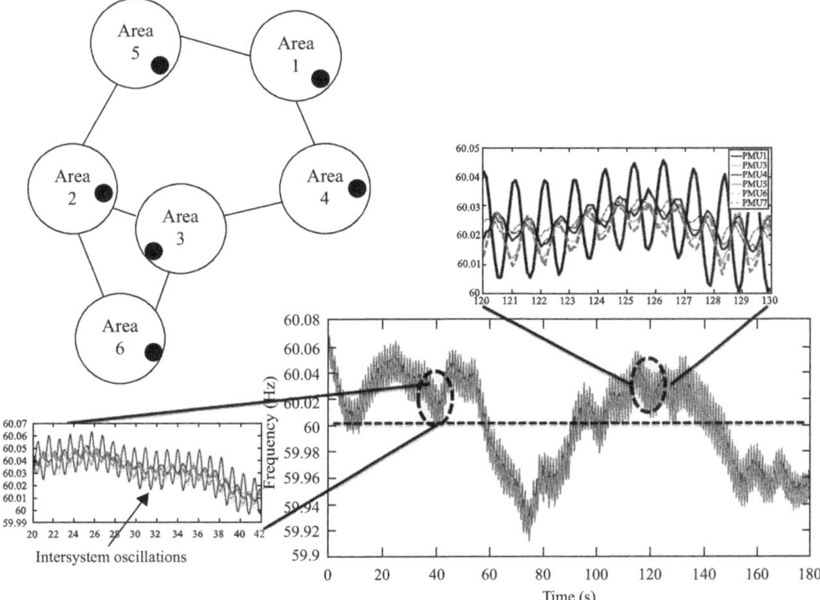

Figure 4.8 Schematic of the study system showing selected frequency measurements. The inset plots show the internal inter-area oscillations

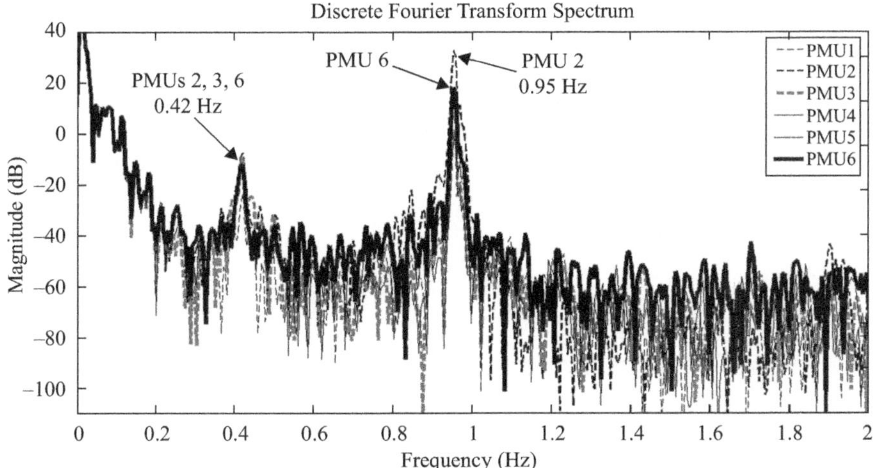

Figure 4.9 Power spectra of PMU measurements

Analysis of the power spectra in Figure 4.9 discloses the presence of a dominant mode at about 0.95 Hz associated with a local mode. A second (inter-area) mode at about 0.42 Hz is also observed.

4.6.1 HHT analysis

As a first step toward the application of the method, the frequency traces were nonlinearly detrended using the EMD-detrending technique in section 4.3.1.

For illustration, analysis of the frequency deviations in Area 3 (PMU 3) was considered. This is the signal with the largest peak-to-peak deviation in Figure 4.8. Figure 4.10 shows the first three IMFs extracted using EMD.

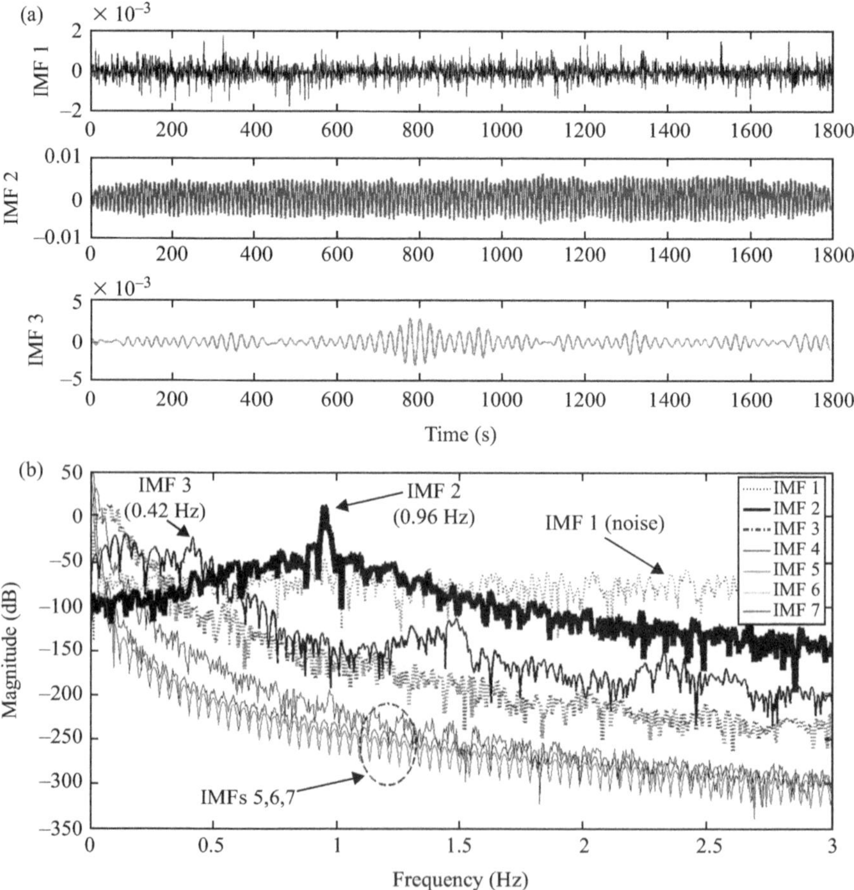

Figure 4.10 Extracted IMFs and their associated spectra: (a) the first extracted IMFs using EMD. (note the different scales); (b) spectra of the IMFs

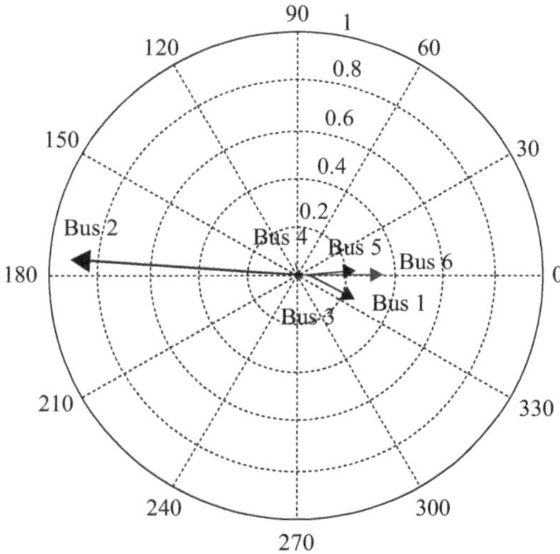

Figure 4.11 Mode shape for the 1.0 Hz mode

The analysis of the power spectra in Figure 4.10b shows a dominant component at about 1.0 Hz.

IMF 1 captures essentially noise, while IMFs 2 and 3 can be seen in Figure 4.10b to characterize the temporal behavior of the 0.96 Hz and 0.42 Hz modal components in Figure 4.10a. Note the magnitude of IMFs 1 and 3, relative to IMF 2.

In turn, the analysis of the complex mode shape for the 1.0 Hz mode in Figure 4.11 shows that bus 2 swings out of phase with buses 1, 3, 5, and 6. Bus 4 shows a less defined oscillation.

Figure 4.12 shows the global power spectrum of PMUs 2, 3, and 6 obtained using the HHT procedure. It can be shown that wavelet analysis leads to similar results. The analysis gives the spatiotemporal representation of modal behavior.

Guided by the mode shape information in Figure 4.12, detailed studies were conducted to examine the instantaneous phase evolution for the 1.0 Hz mode. Figure 4.13 shows the relative extracted instantaneous phases obtained from application of the HHT procedure.

The analysis procedure consists of two main steps:

1. Extract from each PMU measurement the instantaneous phase $\varphi_j(t)$.
2. Compute phases relative to an arbitrary reference, $\varphi_{ref}(t)$.

In this analysis that follows, the instantaneous phase is calculated as

$$\varphi_j(t) = \arctan(x_{Hj}(t)/x_j(t))$$

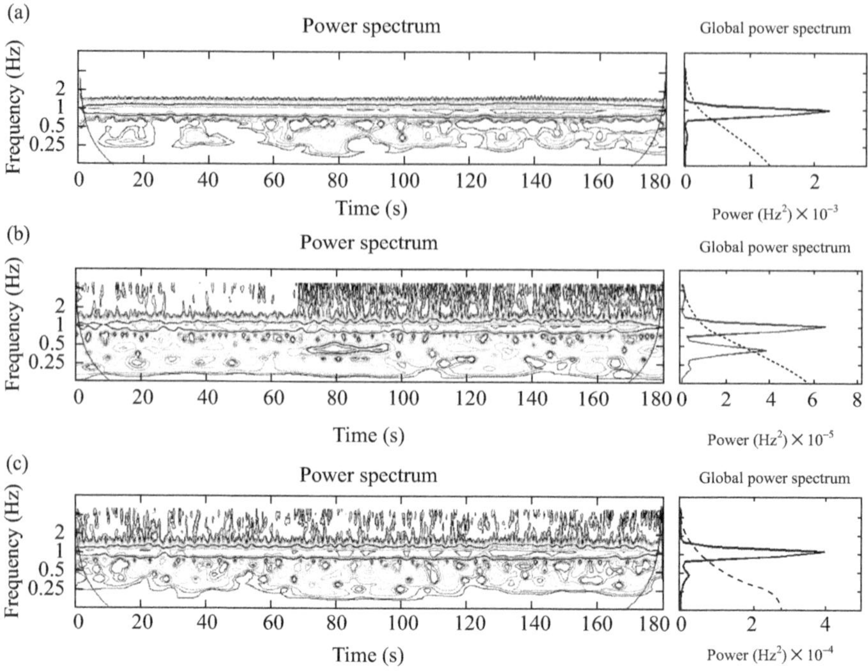

Figure 4.12 Power spectra of frequency measurements in Figure 4.6: (a) PMU 2; (b) PMU 3; (c) PMU 6

For purposes of comparison with conventional modal shapes in Figure 4.11, values are expressed relative to PMU 2 as

$$\hat{\theta}_{PMU_j} = \varphi_{PMU_j} - \varphi_{PMU_2}, \quad j = 1, \ldots, 6, \quad j \neq 2$$

Careful inspection of modal results in Figure 4.13 shows that, for the 0.96 Hz modes, PMUs 1 and 3 swing 180° out of phase with measurements at PMUs 2, 4, 5, and 6. These results are consistent with mode shape information in Figure 4.11, but the results are more general.

Numerical experience with the analysis of complex oscillations shows that change points in the instantaneous phase may signal changes in system behavior. An interesting example is provided in [8] that discusses the impact of switching actions and topological changes on mode shape information.

4.6.2 Wavelet analysis

The approach in section 4.3.2 was used to compute the wavelet decomposition of the measured signals. Using this approach, the measured signals were decomposed as

$$x(t) = \sum_{j=1}^{J} d_j(t) + d_J(t)$$

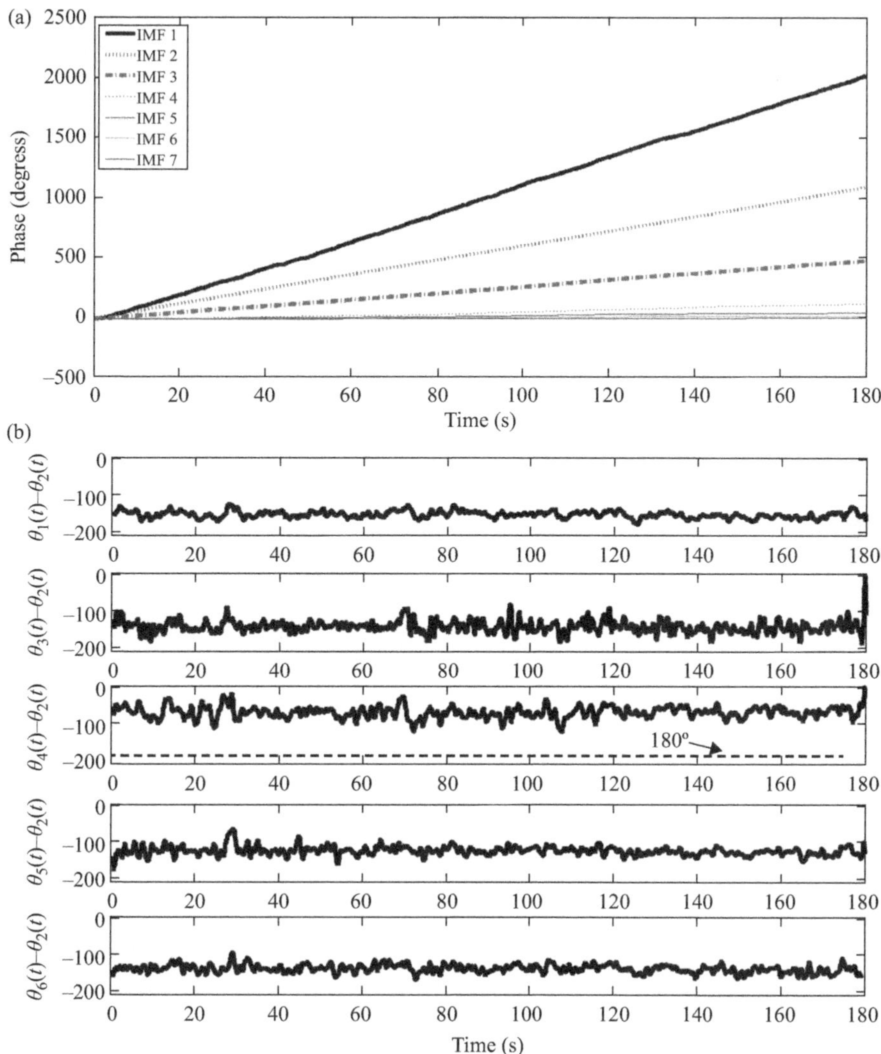

Figure 4.13 Temporal evolution of angle differences between frequency signals. The dashed line shows the reference 180° angle. (a) absolute phases; (b) relative phases

Figure 4.14a shows the extracted dominant modal components (waves 8–11). For comparison, the dominant IMF was identified using Hilbert analysis. The corresponding spectra is shown in Figure 4.14b.

The findings suggest that wavelet and HHT analysis result in a similar decomposition of the signal. In fact, when the mother wavelet is orthogonal, HHT analysis and wavelet analysis results tend to agree for nearly linear oscillations.

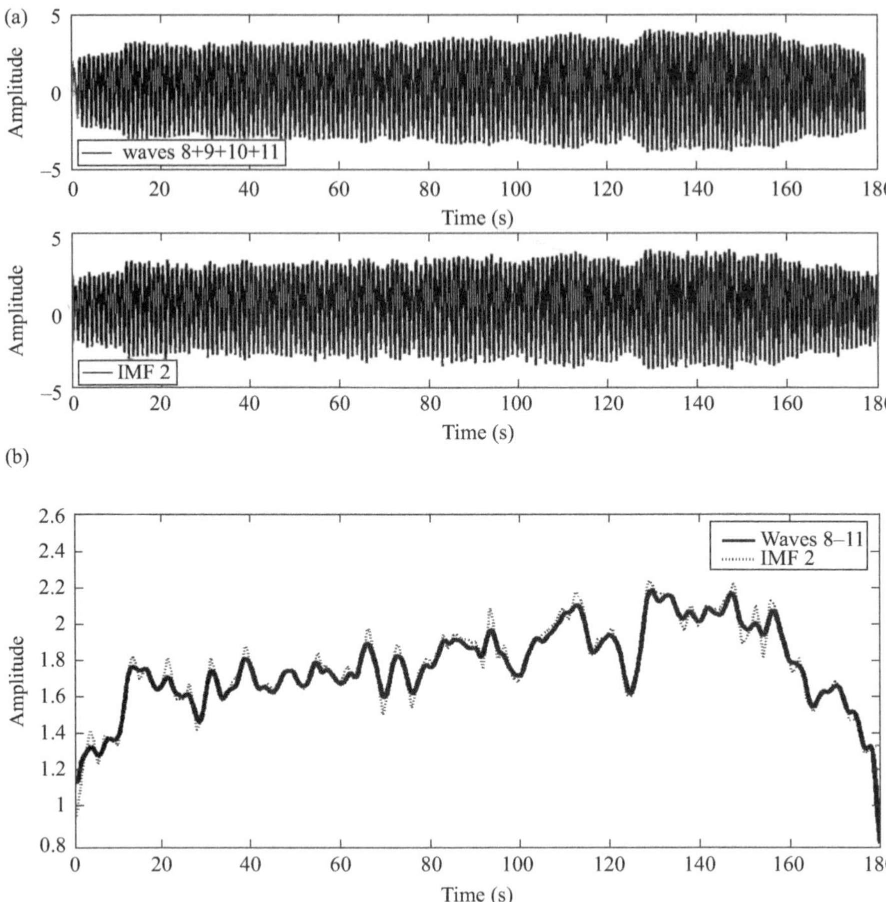

Figure 4.14 Comparison of instantaneous energies from HHT and wavelet analysis: (a) extracted modal components; (b) Fourier spectra of time traces

References

1. John F. Hauer, William A. Mittelstadt, Kenneth. E. Martin, James W. Burns, Harry Lee, John W. Pierre, Daniel J. Trudnowski, 'Use of the WECC WAMS in wide-area probing tests for validation of system performance modeling', *IEEE Transactions on Power Systems*, vol. 24, no. 1, February 2009, pp. 250–257.
2. Power System Dynamic Performance Committee, Task Force on Identification of Electromechanical Modes, Chair: Juan J. Sánchez Gasca, 'Identification of electromechanical modes in power systems', IEEE/PES Special Publication TP462, June 2012.

3. Arturo R. Messina (ed.), *Inter-area Oscillations in Power Systems, A Nonlinear and Nonstationary Perspective*, Springer, New York, NY, 2009.
4. Daniel J. Trudnowski, John W. Pierre, Ning Zhou, John F. Hauer, Manu Parashar, 'Performance of three mode-meter block-processing algorithms for automated dynamic stability assessment', *IEEE Transactions on Power Systems*, vol. 23, no. 2, May 2008, pp. 680–690.
5. A. R. Messina, V. Vittal, Gerald T. Heydt, Timothy T. Browne, 'Nonstationary approaches to trend identification and denoising of measured power system oscillations', *IEEE Transactions on Power Systems*, vol. 24, no. 4, 2009, pp. 1798–1807.
6. N. E. Huang, Z. Shen, S. R. Long, M. C. Wu, H. H. Shih, Q. Zheng, N. C. Yen ... Liu, H. H. (1998). 'The empirical mode decomposition and the Hilbert spectrum for nonlinear and nonstationary time series analysis', *Proceedings of the Royal Society of London A*, vol. 454, 1971, pp. 903–995.
7. A. R. Messina, V. Vittal, 'Non-linear, nonstationary analysis of inter-area oscillations via Hilbert spectral analysis', *IEEE Transactions on Power Systems*, vol. 21, no. 3, 2006, pp. 1234–1241.
8. A. R. Messina, V. Vittal, D. Ruiz-Vega, G. Enríquez Harper, 'Interpretation and visualization of wide-area PMU measurements using Hilbert analysis', *IEEE Transactions on Power Systems*, vol. 21, no. 4, 2006, pp. 1763–1771.
9. Dina S. Laila, Arturo R. Messina, Bikash C. Pal, 'A refined Hilbert–Huang transform with application to interarea oscillation monitoring', *IEEE Transactions on Power Systems*, vol. 24, no. 2, May 2009, pp. 610–620.
10. N. Senroy, S. Suryanarayanan, P. F. Ribeiro, 'An improved Hilbert–Huang method for analysis of time-varying waveforms in power quality', *IEEE Transactions on Power Systems*, vol. 22, no. 4, 2007, pp. 1843–1850.
11. Y. S. Lee, S. Tsakirtzis, A. F. Vakakis, D. M. McFarland, L. A. Bergman, 'Physics-based foundation for empirical mode decomposition', *AIAA Journal*, vol. 47, 2009, pp. 2938–2963.
12. Azadeh Moghtaderi, Patrick Flandrin, Pierre Borgnat, 'Trend filtering via empirical mode decompositions', *Computational Statistics and Data Analysis*, vol. 58, 2013, pp. 114–126.
13. Michael Feldman, *Hilbert Transform Applications in Mechanical Vibration*, John Wiley & Sons, Chichester, West Sussex, 2011.
14. Baojia Chen, Zhenghia He, Xuefen Chen, Honrui Cao, Gaigai Cao, Yangyan Zi, 'A demodulating approach based on local mean decomposition and its applications in mechanical fault diagnosis', *Measurement Science and Technology*, vol. 22, 2011, pp. 1–13.
15. Ingrid Daubechies, Jianfeng Lu, Hau Tieng-Wu, 'Synchrosqueezed wavelet transforms: An empirical mode-like tool', *Applied and Computational Harmonic Analysis*, vol. 30, 2011, pp. 243–261.
16. Jonathan S. Smith, 'The local mean decomposition and its application to EEG perception data', *Journal of the Royal Society*, vol. 2, no. 5, December 2005, pp. 443–454.

17. Davide Lauria, Cosimo Pisani, 'On Hilbert transform methods for low frequency oscillations detection', *IET Generation, Transmission & Distribution*, vol. 8, no. 6, 2014, pp. 1061–1074.
18. Nilanjan Senroy, 'Generator coherency using the Hilbert–Huang transform', *IEEE Transactions on Power Systems*, vol. 23, no. 4, November 2008, pp. 1701–1708.
19. Jose L. Rueda, Carlos A. Juárez, István Erlich, 'Wavelet-based analysis of power system low-frequency electromechanical oscillations', *IEEE Transactions on Power Systems*, vol. 26, no. 3, August 2011, pp. 1733–1743.
20. Jukka Turunen, Jegatheeswaran Thambirajah, Mats Larsson, Bikash C. Pal, Nina F. Thornhill, Liisa C. Harla, William W. Hung, … Tuom as Rauhala, 'Comparison of three electromechanical oscillation damping estimation methods', *IEEE Transactions on Power Systems*, vol. 26, no. 4, 2011, pp. 2398–2407.
21. Samir Avdaković, Elvisa Bećirović, Amir Nuhanović, and Mirza Kušljugić, 'Generator coherency using the wavelet phase difference approach', *IEEE Transactions on Power Systems*, vol. 29, no. 1, January 2014, pp. 271–278.
22. Lijuan Wang, Megan McCullough, Ahsan Kareem, 'Modelling and simulation of nonstationary processes utilizing wavelet and Hilbert transforms', *Journal of Engineering Mechanics*, vol. 140, no. 2, February 2014, pp. 345–360.
23. Martyna Marczak, Victor Gómez, 'Cyclicality of real wages in the USA and Germany: New insights from wavelet analysis', *Economic Modeling*, vol. 47, 2015, pp. 40–52.
24. S. Olhede, A. T. Walden, 'The Hilbert spectrum via wavelet projections', *Proceedings of the Royal Society of London A*, vol. 460, 2004, pp. 955–975.
25. E. Barocio, Bikash C. Pal, A. R. Messina, 'Real-time monitoring as enabler for smart transmission grids', IEEE Power Engineering Society General Meeting, 2011.
26. I. Kamwa, A. Pradhan, G. Joss, 'Robust detection and analysis of power system oscillations using Teager-Kaiser energy operator', *IEEE Transactions on Power Systems*, vol. 26, no. 1, 2011, pp. 323–333.
27. Edward Palmer, 'Nonlinear effects on modal estimates obtained from power system ringdowns', 2011 IEEE Power and Energy Society, General Meeting, San Diego, CA, 2011.
28. Petros Maragos, Thomas F. Quartieri, James F. Kaiser, 'Speech nonlinearities, modulations, and energy operators', 1991.
29. Diego J. Pedregal, Peter C. Young, 'Modulated cycles, an approach to modeling periodic components from rapidly sampled data', *International Journal of Forecasting*, vol. 22, 2006, pp. 181–194.
30. Armando J. Zavala, Arturo R. Messina, 'A dynamic harmonic regression approach to power system modal identification and prediction', *Electric Power Components and Systems*, vol. 42, no. 13, 2014, pp. 1474–1483.
31. Simmo Sarkka, *Bayesian Filtering and Smoothing*, Cambridge University Press, New York, NY, 2013.

32. D. J. Trudnowski, J. M. Johnson, J. F. Hauer, 'Making Prony analysis more accurate using multiple signals', *IEEE Transactions on Power Systems*, vol. 14, no. 1, February 1999, pp. 226–231.
33. Yoshihiko Susuki, Igor Mezic, 'Nonlinear Koopman modes and coherency identification of coupled swing dynamics', *IEEE Transactions on Power Systems*, vol. 26, no. 4, November 2011, pp. 1894–1904.
34. Yoshihiko Susuki, Igor Mezic, 'Nonlinear Koopman modes and power system stability assessment without models', *IEEE Transactions on Power Systems*, vol. 29, no. 2, March 2014, pp. 899–907.
35 E. Barocio, Bikash C. Pal, Nina F. Thornhill, A. R. Messina, 'A dynamic mode decomposition framework for global power system oscillation analysis', accepted for publication in the *IEEE Trans. on Power Systems*, available online: http://ieeexplore.ieee.org/.
36. Luke Dosiek, John W. Pierre, 'Estimating electromechanical modes and mode shapes using the multichannel ARMAX model', *IEEE Transactions on Power Systems*, vol. 28, no. 2, May 2013, pp. 1950–1959.
37. N. Zhou, J. W. Pierre, D. J. Trudnowski, R. T. Guttromson, 'Robust RLS methods for online estimation of power system electromechanical modes', *IEEE Transactions on Power Systems*, vol. 22, no. 3, August 2007, pp. 1240–1249.
38. I. Moreno, A. R. Messina, 'Adaptive tracking of system oscillatory modes using an extended RLS algorithm', *Electric Power Systems Research*, vol. 114, 2014, pp. 28–38.
39. Laure Zanna, Eli Tzipeman, 'Optimal surface excitation of the thermohaline circulation', *Journal of Physical Oceanography*, vol. 38, 2008, pp. 1820–1830.
40. J. Fontane, P. Brancher, D. Fabre, 'Stochastic forcing of the Lamb–Oseen vortex', *Journal of Fluid Mechanics*, vol. 613, 2008, pp. 233–254.
41. Brian F. Farrell and Petros Ioannou, 'Generalized stability theory. Part I: Autonomous operators', *Journal of Atmospheric Sciences*, vol. 53, no. 14, 1996, pp. 2025–2040.
42. Guy-Bart Stan, *Modelling in Biology,* Course notes, June 2014.

Chapter 5
Multisensor multitemporal data fusion

5.1 Introduction

Power system data are multiscale and multivariate in nature. The increasing availability of wide-area measurement systems capable of producing large amounts of multidimensional data has made the use of multivariate data analysis methods more common place.

The preceding chapters described methods for analyzing multisensory multitemporal sensor data. Modern wide-area measurement systems rely on data assimilation to estimate initial and boundary data, to interpolate or smooth sparse or noisy observations, and to evaluate observing systems and dynamical models. The realization of practical data assimilations systems, however, is challenging due to both the high dimensionality of the data and the communication and computational requirements.

Multivariate processes arise when several related time series processes are observed simultaneously over time instead of observing just a single series [1]. Existing wide-area monitoring systems (WAMS) architectures provide only partial state information; as a result, the information provided by individual sensors is incomplete, inaccurate, and/or unreliable.

This chapter examines the feasibility of using multisensor data fusion techniques for monitoring and analyzing power system oscillatory behavior. A common conceptual mathematical framework for integrating multiscale data to improve situational awareness is provided. The framework includes techniques to classify and extract dynamic patterns from multisensor multiscale data. Outlier detection and methods to evaluate the statistical significance of the results obtained from the different methods are also discussed.

The methods are implemented and compared in terms of their ability to fuse data from multiple sensors.

5.2 Data fusion principles

Data fusion can be broadly defined as the process of combining data from different sources (sensors) to provide a robust and complete description of a process of interest [2–4]. Sensors may include phasor measurement units (PMUs), dynamic frequency recorders, relay-based PMUs, and other sensors. At the core of these systems are advanced statistical and mathematical techniques used to process vast amounts of data in near real time.

Integrating complex dynamic data from different sensors is a challenging problem due to both communication and computational issues [4]. As noted earlier, observations may be noisy, be heterogeneous, and exhibit differing spatial and temporal characteristics that make the identification of critical system features difficult [2, 5–7].

The application of data fusion techniques has been advanced greatly by the development of WAMS. Figure 5.1 provides a schematic representation of the processing chain for an individual sensor [8]. Sensor placement is critical for true intelligence in data fusion techniques (monitoring). Typically, the sensor signal is processed and used to formulate some decision about the system or to make a prediction about the future behavior of the physical system under consideration [9].

There are numerous methods and architectures for multisensor data fusion, and different applications may require fusion at different levels. A classification of data fusion technique is given in recent work [10, 11]. Structures used in multisensor data fusion systems may be of the centralized and distributed types. Distributed fusion architectures are of special interest for power swing monitoring due to various reasons: (a) there is a requirement for information at regional as well as global scales, (b) fusion architectures can be integrated to existing local power data concentrators, and (c) sensors in the distributed fusion structure can be independent from each other and potentially heterogeneous.

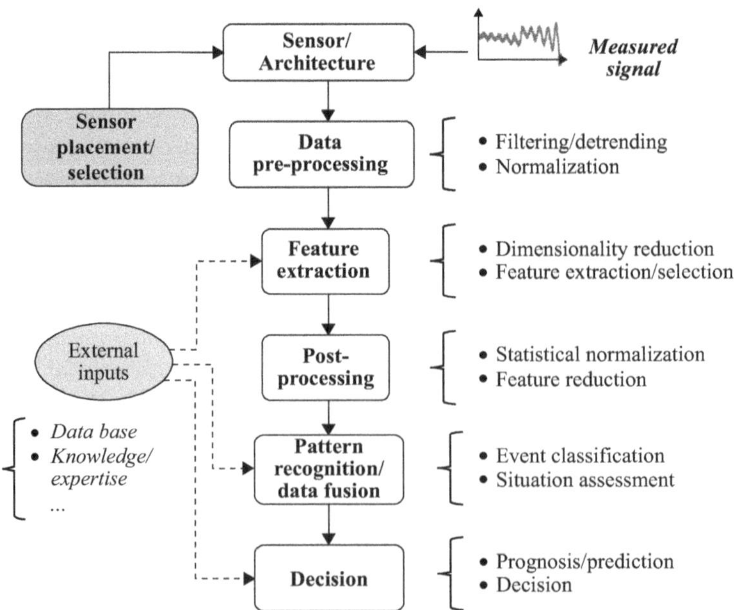

Figure 5.1 Processing chain for sensor data fusion

Drawing on previous research on system health monitoring, a data fusion framework that combines ideas from time–frequency feature extraction techniques with statistical approaches is proposed in this chapter.

Figure 5.2 shows a schematic of the adopted data fusion architecture. The system consists of several fusion centers integrated to regional phasor data concentrators. A hierarchical architecture is adopted in which local measurements at the regional fusion centers are transmitted to a higher level fusion system (a global power data concentrator) where these measurements are fused.

The model consists of four major modules or levels [11, 12]:

1. Data acquisition and cleansing (pre-processing)
2. Feature extraction and feature selection
3. Feature-level fusion
4. Decision support

These processing steps are presented schematically in Figure 5.1.

Pre-processing may include data alignment and association, dimensionality reduction, filtering, and detrending. PMU data is time synchronized but may require filtering and detrending at local or feature levels.

At the second stage, feature-extraction level, key damage-sensitive properties are extracted and classified. Typically, these include modal features and measures of energy distribution, such as mode shapes, but may include more general pattern

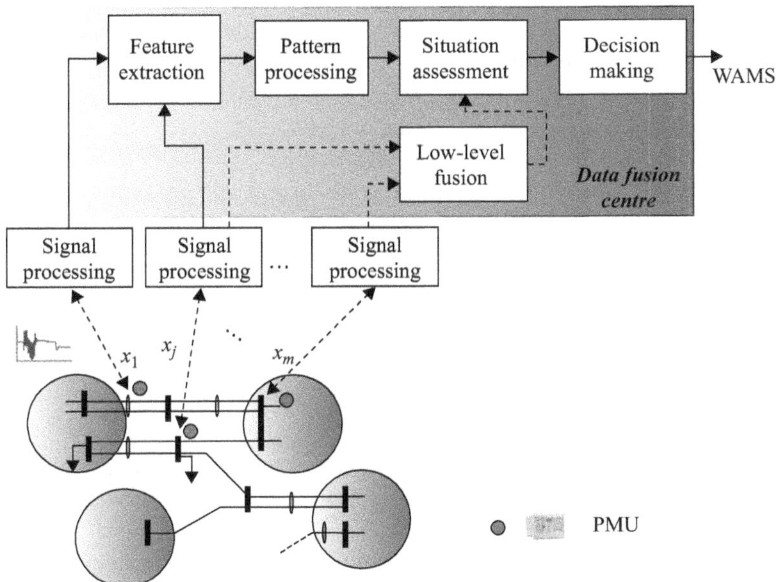

Figure 5.2 General architecture of a data fusion scheme showing the data fusion engine

recognition techniques. Two different types of data information can be distinguished: raw data provided by multiple sensors and feature-level data from a feature-extraction technique.

Emerging from the feature extraction module are feature vectors. Pattern recognition techniques can then be used to identify features from the transient response as well as to discern the significant dynamic patterns containing dominant features in data.

In the decision level, the output of the data fusion technique is assimilated into a decision support framework based on selected relevant features. The decision model may combine selected outputs from the feature extraction module with offline criteria or information from other sources, such as supervisory control and data acquisition (SCADA) systems.

Outputs of the regional fusion centers can be used for damage detection and location, and health monitoring of the system using existing wide-area monitoring, protection, and control systems.

Figure 5.2 presents an abstract architecture that has both online and offline functions. It consists of the following main components: a sensor subsystem, a data fusion engine subsystem, and an application subsystem.

In the following sections, the roles of monitoring and sensor technologies in the development of a practical data fusion system will be examined.

5.3 Data pre-processing and transformation

Measured data are high-dimensional, noisy, and nonstationary in nature. Changes in the dynamical properties of the signals may affect the performance and applicability of the signal processing or data fusion methods. Thus, for instance, the covariance matrices calculated from incomplete data may not be positive definite and produce negative eigenvalues. This, in turn, can affect the application of methods that rely on eigenvalue decomposition [13].

The pre-processing stage is aimed at cleansing the measured signals and reducing the dimension of the data vector to eliminate as much redundancy as possible. Data preprocessing may also be necessary for optimal algorithm performance [14].

Typical procedures at local (sensor) level include data normalization, noise reduction, trend analysis, and rejection of outliers.

5.3.1 Bandpass filtering and denoising

In practice, signature analysis can be applied to a subset of modes (a frequency band) to uniquely characterize the time evolution of critical components giving rise to the observed oscillations.

In order to gain a better understanding of the nature of this problem, consider m variables $x_j(t)$, $j = 1, \ldots, m,$ which might represent measured data at m PMUs or other recording devices. For simplicity and clarity, a centralized data fusion structure is considered first. The more general and interesting case is discussed in section 5.4.

Let the measured variables be observed at N times, $t = t_1, t_2, \ldots, t_N$. Expanding $x_j(t)$ in terms of modal components yields

$$\begin{aligned} x_1(t) &= c_{11}(t) + c_{12}(t) + \ldots + c_{1p_1}(t) + r_{p_1}(t) \\ x_2(t) &= c_{21}(t) + c_{22}(t) + \ldots + c_{2p_2}(t) + r_{p_2}(t) \\ &\vdots \\ x_m(t) &= c_{m1}(t) + c_{m2}(t) + \ldots + c_{m_{pm}}(t) + r_{p_m}(t) \end{aligned} \tag{5.1}$$

where, in general, $c_{kj}(t) = A_{kj}(t)\cos(\varphi_{kj}(t))$ with associated amplitudes and phases $A_{kj}(t)$, $\dot{\varphi}_{kj}(t) > 0$, $\forall t$, and the r_{p_k}, $k = 1, \ldots, m$ are the number of relevant modes captured by each sensor.

It is noted that this model is general and could represent various modal decompositions such as Prony, Hilbert, or wavelet decompositions.

Several remarks are now in order about this model.

- Depending on the signal processing technique employed, each set $\{c_{k1}(t), c_{k2}(t), \ldots, c_{kp_k}(t)\}$ corresponds to a given frequency $f_k = d\varphi_{kj}/dt$ or a frequency band $[f_{k\min}, f_{k\max}]$.
- This information can be arranged into feature vectors $\mathbf{x}_{f_j}(t) = [c_{j1}(t_1) \quad c_{j1}(t_2) \quad \cdots \quad c_{j1}(t_N)], j = 1, \ldots, m$ and used for condition monitoring, assessment, and prediction as discussed below.
- The feature vectors can then be collected into feature matrices of the general form

$$\mathbf{X}_f(t) = [\mathbf{x}_{f_1}(t) \quad \mathbf{x}_{f_2}(t) \quad \cdots \quad \mathbf{x}_{f_m}(t)]$$

Typically, the feature space described by these models is high-dimensional and sparse, which results in data inconsistency, uncertainty in modal estimates, and time-consuming analysis processes.

5.3.2 Local-level fusion

Figure 5.3 shows an elementary representation of the proposed feature-level fusion approach based on the representation in (5.1). More general representations are discussed below. In this approach, raw measurements are decomposed into a set of modal components and used to generate feature-level observational data. Outliers and other artifacts are detected and removed from the data sets using simple statistical models.

Several interpretations are possible as discussed in the following section.

5.4 Feature extraction and feature selection

5.4.1 Feature extraction

Feature extraction is the process of identifying damage-sensing properties from the measured system response [11]. In principle, all various methods discussed in Chapter 4 can be used for feature extraction from system response.

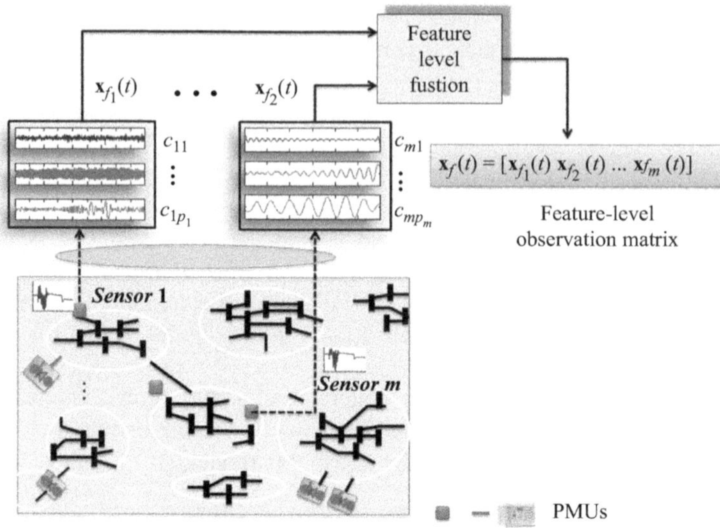

Figure 5.3 Multivariate space-time data fusion

Features of interest may include the mean process at a site or regional level, modal properties, signal's amplitude and energy, or other quantities. Generally speaking, feature extraction may involve some type of model reduction as the original measurements may contain components irrelevant to the problem of interest or be highly correlated. The individual features of interest are usually aggregated and arranged into an n-dimensional feature vector.

Feature selection is the process of selecting those components of a feature vector which carry most of the discriminatory power of the feature. In broad terms feature extraction refers to the process of transforming the existing features into a lower dimensional space. Examples of these techniques include proper orthogonal decomposition (POD), independent component analysis (ICA), and principal component analysis (PCA) [4].

In section 5.4.2 that follows, the issue of data compression for feature extraction is introduced. Other aspects of interest are discussed in Chapter 7.

5.4.2 Data compression

As discussed above, linear (nonlinear) multivariate methods are commonly used to perform data compression prior to the feature extraction process, when data from multiple measurement points are available. This process transforms the time series from multiple measurement points into a single time series, preserving as much of the relevant information as possible during the dimensionality reduction.

One of the important steps in the design of wide-area systems is the selection of the best feature sets representing system behavior. This process is generally data and application dependent.

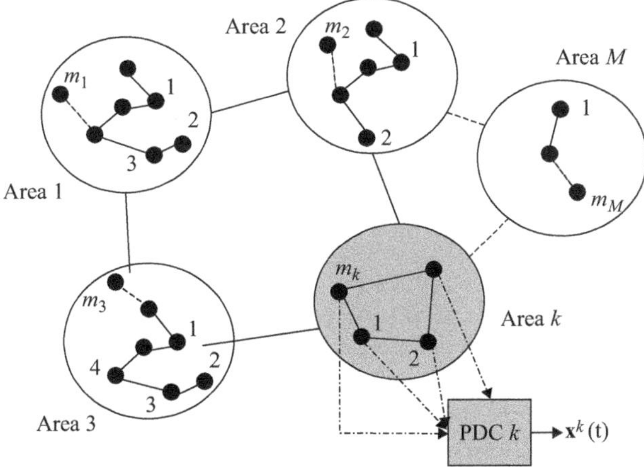

Figure 5.4 Multiarea multisensor power system. The m_k, $k = 1, \ldots, M$, represent the number of sensors associated with each area or local PDC

Two basic processing methodologies are considered in this context:

1. PCA-based data compression
2. Multiway POD

Within the framework of decentralized WAMS structures, data collected from various control centers may be efficiently fused using multisensor fusion techniques.

To introduce the more general ideas that follow, consider a power system composed of M areas, each of which contains a number of sensors. Again, the areas are indexed $\{1, 2, \ldots, M\}$. For clarity of illustration and visualization assume that each area has m_k sensors, $k = 1, \ldots, M$, where it is assumed that $m_1 \neq m_2, \ldots, m_k, \ldots, m_M$. Figure 5.4 provides a conceptual representation of the adopted model.

Using the notation of Chapter 4, let now the time evolution of measured signals in area k be expressed in vector form as

$$\mathbf{x}^k(t) = \begin{bmatrix} x_1^k(t) \\ x_2^k(t) \\ \vdots \\ x_{m_k}^k(t) \end{bmatrix}$$

$$= \underbrace{\begin{bmatrix} c_{11}^k(t) \\ c_{21}^k(t) \\ \vdots \\ \underbrace{c_{m_k 1}^k(t)}_{Mode\ 1} \end{bmatrix} + \cdots + \begin{bmatrix} c_{1j}^k(t) \\ c_{2j}^k(t) \\ \vdots \\ \underbrace{c_{m_k j}^k(t)}_{Mode\ j} \end{bmatrix} + \cdots + \begin{bmatrix} c_{1p_1}^k(t) \\ c_{2p_2}^k(t) \\ \vdots \\ \underbrace{c_{m_k p_k}^k(t)}_{Mode\ p} \end{bmatrix}}_{\text{Oscillatory components}} + \underbrace{\begin{bmatrix} m_1^k(t) + \xi_1^k(t) \\ m_2^k(t) + \xi_2^k(t) \\ \vdots \\ m_{m_k}^k(t) + \xi_{m_k}^k(t) \end{bmatrix}}_{\text{Noise and trends}} \tag{5.2}$$

where the term $c_{ij}^k(t)$, $i = 1, \ldots, p_i$, $j = 1, \ldots, m_k$ on the *rhs* of (5.2) represents oscillatory modal components, and the second term represents noise and trends in the signal. For mathematical convenience in the discussion that follows, it is assumed that $p_1 = p_2 = \ldots = p_k$.

Each column of (5.2) gives the time evolution of a mode or frequency band. In most cases of interest, however, each column in the first *rhs* term of (5.2) can be associated with a specific frequency.

A schematic representation of the model in (5.2) is shown in Figure 5.5. By disregarding unimportant or unphysical components, the above model can be used to represent specific behavior. The problem of interest becomes that of fusing data to extract specific information of interest. Observe that fusion involves the solution of three main problems: (a) extracting from a set of sensors the oscillatory behavior of interest, (b) selecting the most useful information, and (c) fusing data.

In most practical applications, the analysis focuses on a given frequency band, for instance, associated with critical inter-area modes. Three cases are of interest here:

1. The analysis of individual scales associated with a given mode of interest
2. Multiscale analysis associated with a given frequency band, that is, the frequency range associated with local or inter-area modes
3. Monitoring based on the entire modal space (state reconstruction)

From a physical perspective, the response matrix could be constructed in two basic ways: either a single modal component common at each measurement site or each column defined as a collection of modal components.

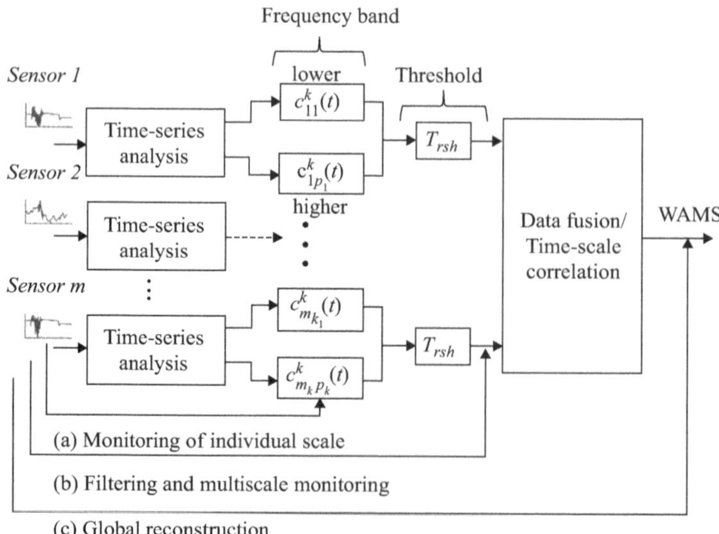

Figure 5.5 Multivariate and multiscale monitoring of system behavior: single area case

To formalize the model, consider the first case above. With reference to (5.2), the modal response matrix for area k corresponding to mode j is defined as

$$\mathbf{X}_j^k(t) = \begin{bmatrix} c_{1j}^k(t_1) & c_{1j}^k(t_2) & \cdots & c_{1j}^k(t_N) \\ c_{2j}^k(t_1) & c_{2j}^k(t_2) & \cdots & c_{2j}^k(t_N) \\ \vdots & \vdots & \ddots & \vdots \\ c_{m_kj}^k(t_1) & c_{m_kj}^k(t_2) & \cdots & c_{m_kj}^k(t_N) \end{bmatrix}, \quad j = 1, \ldots, p_k \tag{5.3}$$

where p_k is the number of modes of interest, and N is the number of snapshots. This leads to a multidimensional or multiblock representation when a number of modes, $j = 1, \ldots, p$, are of interest.

In an analogous manner, the modal response associated with the ith sensor of area k can be defined as

$$\mathbf{X}^k(i,t) = \begin{bmatrix} c_{i1}^k(t_1) & c_{i1}^k(t_2) & \cdots & c_{i1}^k(t_N) \\ c_{i2}^k(t_1) & c_{i2}^k(t_2) & \cdots & c_{i2}^k(t_N) \\ \vdots & \vdots & \ddots & \vdots \\ c_{ip_k}^k(t_1) & c_{ip_k}^k(t_2) & \cdots & c_{ip_k}^k(t_N) \end{bmatrix}, \quad i = 1, \ldots, m_k \tag{5.4}$$

of dimensions $p_i \times N$.

It should be stressed that (5.4) gives the contributions of all modes to the time evolution of a given sensor. Physically, each entry i,j captures the time evolution of the jth mode at sensor i, and time t_j.

Note also that (5.3) and (5.4) could be defined in alternative forms. Any data fusion results in a three-way decomposition, which represents the different signals (modes) and characteristics present in the data as a function of three parameters: sensor location, time, and modal information.

5.4.3 Individual scales

The simplest approach to space–time correlation is to obtain, at each time instant, an average estimate of the spatial distribution of a modal component. Given a set of modal responses of the form (5.4), data fusion at a given modal level can be obtained from the time average of the individual modal estimates. For area k, the average behavior of the jth modal response across the m_k sensors can be written as

$$\hat{c}_{ij}^k(t_j) = \frac{1}{m_k} \sum_{i=1}^{m_k} c_{ij}^k, \quad i = 1, \ldots, p, \quad j = 1, \ldots, N \tag{5.5}$$

as suggested in Figure 5.6. This represents an elementary data fusion approach.

5.4.4 Filtering and multiscale monitoring

The above models extend naturally to the multiscale case. Based on the general ideas in Chapter 2 consider a decentralized WAMS structure consisting of M local

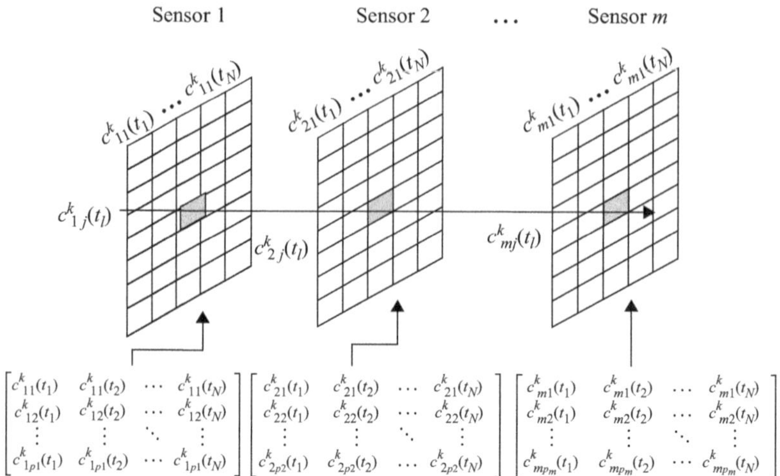

Figure 5.6 Elementary data fusion

PDCs. Here, each PDC processes a set of individual measurements and produces a set of modal matrices of the form (5.3).

When a frequency band associated with a set of dominant inter-area modes is of interest, the modal matrices are of the form

$$\mathbf{X}^k(t) = \mathbf{X}^k_{pdc}(t) = \begin{bmatrix} c_{11}(t) & c_{12}(t) & \cdots & c_{1p_1}(t) \\ c_{21}(t) & c_{22}(t) & \cdots & c_{2p_2}(t) \\ \vdots & \vdots & \ddots & \vdots \\ c_{m1}(t) & c_{m2}(t) & \cdots & c_{mp_m}(t) \end{bmatrix}, \quad k = 1, \ldots, M \tag{5.6}$$

where, for simplicity, it is assumed that $p_1 = p_2 = \ldots = p_m$.

In its simplest form, statistical averaging can be used to obtain the time average estimates of modal behavior at each time instant as suggested in Figure 5.7. When the number of variables is large (i.e., the number of local data concentrators is large and/or the number of modes or the observation period increases), however, direct use of these techniques is not practical, especially when the data sets contain a lot of redundant information.

Such redundancy results from multiple measurements of the same variable or constraining relationships between different variables.

In this case, the modal average estimate can be expressed in the form

$$\mathbf{X}_f(l,t) = \begin{bmatrix} \hat{c}_{11}(t) & \hat{c}_{12}(t) & \cdots & \hat{c}_{1p_1}(t) \\ \hat{c}_{21}(t) & \hat{c}_{22}(t) & \cdots & \hat{c}_{2p_2}(t) \\ \vdots & \vdots & \ddots & \vdots \\ \hat{c}_{m1}(t) & \hat{c}_{m2}(t) & \cdots & \hat{c}_{mp_m}(t) \end{bmatrix}, \quad l = 1, 2, \ldots, m \tag{5.7}$$

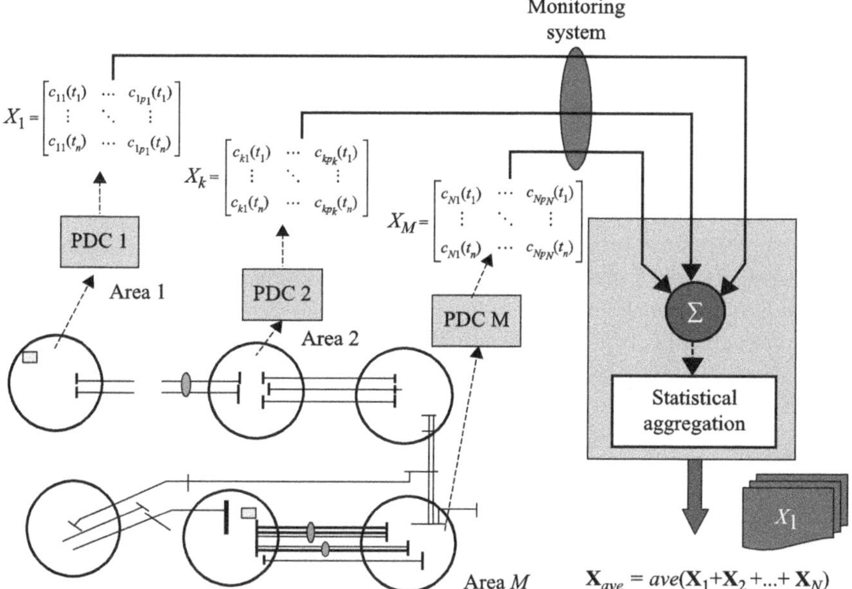

Figure 5.7 Statistical averaging of modal components

where the individual entries, $\hat{c}_{lj}(t)$, are given by

$$\hat{c}_{lj}(t) = \frac{1}{M}\sum_{j=1}^{M} c_{lj}, \quad l = 1, \ldots, m$$

This leads to a model of the form

$$\begin{aligned} \mathbf{X}_f &= \frac{1}{M}\left(\mathbf{X}_{f_1} + \mathbf{X}_{f_2} + \ldots + \mathbf{X}_{f_M}\right) \\ &= \frac{1}{M}\left(\mathbf{X}_f(1,t) + \mathbf{X}_f(2,t) + \ldots + \mathbf{X}_f(M,t)\right) \end{aligned} \tag{5.8}$$

As observed, both central processing unit (CPU) time and memory needed for analyzing large sets of modal decompositions increase rapidly with the size of the system (the number of PDCs) and the observation period. It is possible, however, to overcome this problem by using multisensor data fusion techniques.

5.5 Multisensor fusion methodologies for system monitoring

Data-driven techniques constitute the most straightforward approach to data fusion. These include techniques such as blind source separation (BSS), PCA, POD, canonical correlation analysis (CCA), and independent component analysis (ICA), among others. These methods are particularly attractive in this context since they can achieve useful decompositions of the multimodal or multiset data.

It has been suggested that multisensor data fusion may enhance signal detection. There are several advantages to such an approach. First, raw measurements can be used to the estimate the relative magnitude and phase of the oscillatory modes. Second, these techniques can be used for diagnosis of system disturbances and assessing specific component contribution to system behavior [15].

5.5.1 Single-scale analysis

The single-scale PCA method has been introduced in Chapter 3. PCA finds a lower dimensional subspace that best preserves the data variance, and where the variance of the data is maximal.

Mathematically, PCA transform an $m \times N$ observation matrix, $\mathbf{X}$, by combining the variables as a linear weighted sum as

$$\mathbf{X}_f = \mathbf{P}^T\mathbf{T} + \mathbf{E} = \sum_{j=1}^{p} \mathbf{p}_i^T \mathbf{t}_i + \mathbf{E} = \hat{\mathbf{X}}_f + \mathbf{E} \tag{5.9}$$

where $\mathbf{P}$ is an $m \times N_L$ principal-component loadings, $\mathbf{T}$ is the $m \times N_L$ principal-component scores that represent the contribution of the score variables to the reconstruction of the original process variables, and $\mathbf{E} \in^{m \times N_L}$ is the residual matrix; p is the number of principal components retained in the model.

Particular cases of this model are the model in (3.16), singular value decomposition (SVD)-based POD and PCA. The specifics of the above derivation are presented in detail in statistical texts.

Several features of this model are worth pointing out:

1. The first principal component describes the largest amount of variation in the observation matrix, $\mathbf{X}$. The projected data in the new space is given by $\hat{\mathbf{X}}_f = \mathbf{P}^T\mathbf{T}$.
2. The loading vectors are orthonormal and provide the direction with maximum variability.
3. The scores from the different principal components are the coordinates of the objects in the reduced space.

These techniques are best suited for the analysis of steady-state data containing linear relationships between the variables, and are single-scale in nature. Since these conditions are often not satisfied in practice, several extensions and generalization have been developed [16–18]. These include complex-based analytical formulations, dynamic PCA, and neural network-based PCA.

5.5.2 Nonlinear PCA using auto-associative neural networks

Auto-associative neural networks (AANN) have been recently proposed as an extension to linear PCA analysis [19, 20]. These networks consist of three internal layers, namely the mapping and de-mapping layers and a bottleneck layer in addition to one input and one output layers. The input and output layers have equal number of nodes.

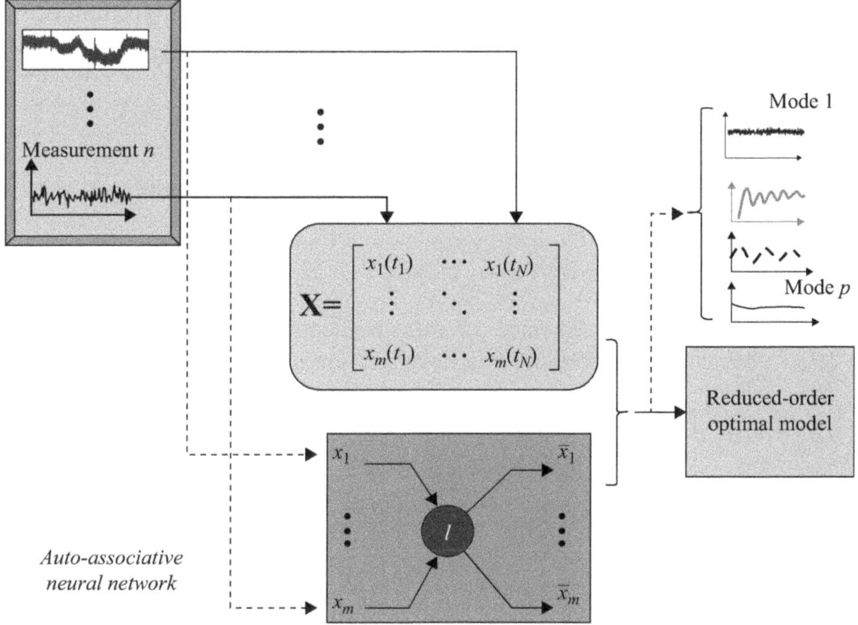

Figure 5.8 Schematic diagram of an AANN as an extension to conventional POD analysis

Figure 5.8 illustrates the AANN architecture used in this analysis. As shown in this plot, the AANN is equivalent to the POD procedure, where the inputs to the AANN are the raw measurements and the outputs are the modal components $c_1(t)$, $c_2(t)$, ..., $c_m(t)$.

The key advantage of this approach is that the POD analysis extends naturally to the nonlinear framework [21] and reduces the dimensionality of the ensemble of measurements much better than does linear PCA.

In what follows, extensions to the above model are provided in the context of PCA analysis, but the discussion applies to similar techniques described above.

5.5.3 Multiblock POD (PCA) analysis

The POD problem can be extended to multiple data sets using the framework developed in [19, 20]. Several hybrid multiscale PCA analysis methods based on combining PCA and time-frequency analysis techniques, namely wavelet and HHT transform and Kalman analysis, have developed. Due to its multiscale nature, these models are appropriate for treating multiscale, multivariate data from multiple sensors.

These techniques combine the idea of PCA to decorrelate the variables by extracting a linear relatonship with that of time-series analysis to extract deterministic features and approximately decorrelate autocorrelated measurements.

The WAMS architecture considered here is of the distributed type described in Chapter 2. Again, to formalize the model, consider a network of m sensors. Three basic types of applications, involving ways to unfold the data, can be considered:

1. Raw data from multiple synchrophasors
2. Temporal scales from a time-series analysis technique
3. Raw data (temporal scale) from a window-based analysis technique

Discussion of application 3 is postponed until Chapter 7.

5.5.3.1 Raw-level data

Consider the case of a sequence of measurements $\{x_i^k(t_1), x_i^k(t_2), \ldots, x_i^k(t_N)\}$, $i = 1, \ldots, m_k$. One can define the raw-level feature matrix for area k as follows:

$$\mathbf{X}_{raw}^k(t) = \begin{bmatrix} x_1^k(t_1) & x_1^k(t_2) & \cdots & x_1^k(t_N) \\ x_2^k(t_1) & x_2^k(t_2) & \cdots & x_2^k(t_N) \\ \vdots & \vdots & \ddots & \vdots \\ x_{m_k}^k(t_1) & x_{m_k}^k(t_2) & \cdots & x_{m_k}^k(t_N) \end{bmatrix}, \quad k = 1, \ldots, M$$

By preserving the number of samples, N, the feature matrix can now be rewritten in the form

$$\mathbf{X}_f = \begin{bmatrix} \mathbf{X}_{raw}^1(t) \\ \hdashline \mathbf{X}_{raw}^2(t) \\ \hdashline \vdots \\ \hdashline \mathbf{X}_{raw}^M(t) \end{bmatrix}$$

In practice, however, direct analysis of the feature matrix, $\mathbf{X}_f$, may become infeasible for a multisensor multiarea power system. A better alternative is to adopt a multiblock statistical analysis approach. This achieves two things: (a) the computational burden is reduced, and (b) the analysis of correlations between adjacent areas is obtained.

Figure 5.9 illustrates schematically the suggested approach. The analysis procedure divides into three principal phases:

1. Assembling the individual data matrices
2. Unfolding of the data
3. Extracting relationships between and within the sets of blocks, for instance, associated with data from different regions or control centers.

5.5.3.2 Sensor level

A first application of interest focuses on the analysis of raw data at a local level. To introduce this notion, let the observed system response at the jth sensor be expressed as

$$x_j^k(t) = c_{j1}^k(t) + c_{j2}^k(t) + \cdots + c_{jp_j}^k(t) + m_j^k(t) + \xi_j^k(t)$$

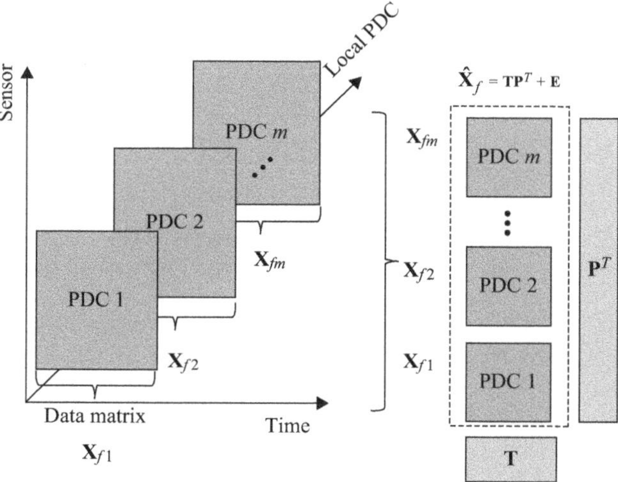

Figure 5.9 Hybrid multiblock PCA analysis resulting in a three-way decomposition of measured data: raw-level data fusion

One can therefore define a feature-level matrix as

$$\mathbf{X}^k(j,t) = \begin{bmatrix} c_{j1}^k(t_1) & c_{j1}^k(t_2) & \cdots & c_{j1}^k(t_N) \\ c_{j2}^k(t_1) & c_{j2}^k(t_2) & \cdots & c_{j2}^k(t_N) \\ \vdots & \vdots & \ddots & \vdots \\ c_{jp_j}^k(t_1) & c_{jp_j}^k(t_2) & \cdots & c_{jp_j}^k(t_N) \end{bmatrix}, \quad j = 1, \ldots, m_k$$

of dimension $p_j \times N$ where p_j is the number of modal components of interest, m is the number of sensors, and N is the number of snapshots. Assuming that the modes of interest, p_j, is the same for all sensors, this yields a set of m_k blocks of data that must be processed together.

The analysis can be readily extended to the study of multiarea power systems as suggested in Figure 5.10. The analysis, however, may soon become infeasible as the number of areas (and the associated PMUs) increases.

Conventionally, the local observation matrices $\mathbf{X}_{f_j}, j = 1, \ldots, M$, coming from the various PDCs are converted to a two-dimensional data matrix by unfolding the data. This process results in an extended matrix of the form

$$\mathbf{X}_f = [\mathbf{X}_{f_1} \quad \mathbf{X}_{f_2} \quad \cdots \quad \mathbf{X}_{f_m}]^T \tag{5.10}$$

of dimension $p \times mN$ where p is the number of modal components of interest, m is the number of sensors, and N is the number of snapshots.

In the final step, the unfolded matrix is decomposed as

$$\hat{\mathbf{X}}_f = \mathbf{TP}^T + \mathbf{E} \tag{5.11}$$

where $\mathbf{T}$ is the scores matrix, $\mathbf{P}$ is the loading matrix, and $\mathbf{E}$ is the residual vector.

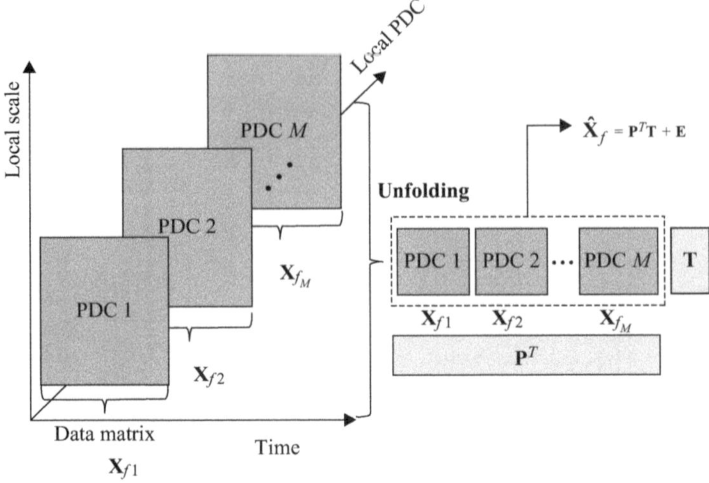

Figure 5.10 Hybrid multiblock PCA analysis

The approach is especially suitable to detect changing patterns between data in different PDCs and detect disturbances on a global scale.

5.5.3.3 Analysis of temporal scales

A second aspect of interest in wide-area monitoring is the analysis of modal behavior at different temporal scales. Rather than focusing on a set of sensors, the aim is to analyze the combined effect of a set of modal components on system behavior.

Consider, to this end, the case of a single mode of interest, say the jth mode. From (5.1), the contribution of the jth modal response to the observed response is given by

$$\mathbf{x}_j^k(t) = \underbrace{\begin{bmatrix} c_{1j}^k(t) \\ c_{2j}^k(t) \\ \vdots \\ \underbrace{c_{m_kj}^k(t)}_{Mode\ j} \end{bmatrix}}_{\text{Oscillatory components}}$$

It then follows that the observation matrix can be rewritten in the form

$$\mathbf{X}_j^k(t) = \begin{bmatrix} c_{1j}^k(t_1) & c_{1j}^k(t_2) & \cdots & c_{1j}^k(t_N) \\ c_{2j}^k(t_1) & c_{2j}^k(t_2) & \cdots & c_{2j}^k(t_N) \\ \vdots & \vdots & \ddots & \vdots \\ c_{m_kj}^k(t_1) & c_{m_kj}^k(t_2) & \cdots & c_{m_kj}^k(t_N) \end{bmatrix}, \quad j = 1, \ldots, p_k$$

As the number of sensors in each area of the system is assumed to be different, that is, $m_1 \neq m_2, \ldots, m_k, \ldots, m_M$, a modification to the single-sensor case is needed.

Table 5.1 summarizes the characteristics of some of these formulations.

It should be emphasized that conventional two-block representations may not be well suited to the analysis of common features or relationships between blocks. One such possible generalization is the multiblock partial least-squares analysis.

5.5.3.4 Multiscale PCA

These techniques combine the ability of time–frequency analysis techniques such as HHT or wavelets to extract deterministic features with that of PCA/POD to extract the cross-correlation or relationship between variables to separate deterministic features from stochastic processes. Due to its multiscale nature, multiscale PCA is appropriate for the modeling of processes containing contributions from dynamic events whose behavior changes over time and frequency.

The process can be summarized as follows:

1. Decompose power system measured data into its contributions in different regions of the time–frequency space, $\{c_{k1}(t), c_{k2}(t), \ldots, c_{kp_k}(t)\}$.
2. Construct the expanded data matrix (5.3) or (5.4) and build a multiblock representation of the system.
3. Perform multiscale PCA analysis on the multiblock representation including relevant scales.
4. Reconstruct the original data matrix from the selected frequency bands.

The method allows the nonstationary behavior of dynamic profile to be analyzed into separate frequency bands that can facilitate the interpretation of intersystem oscillations. Table 5.2 describes recent work on this subject.

Table 5.1 Block system representations

Type	Description	Dimension
Sensor	Local sensor	$p_k \times N \times m_k$
PDC	Local area	$m_k \times N_L \times N_L$
Global	Multi-area	$m_k \times p \times N_L$

Table 5.2 Multiscale PCA formulations

Type	Description
HHT-based MSPCA [22]	Local decomposition of individual measurements using EMD followed by PCA
Wavelet-based MSPCA [23]	Local decomposition of individual measurements using wavelets followed by PCA
Kalman-based PCA [24]	Kalman filter combined with PCA analysis

An efficient visualization technique becomes indispensable at this stage. Discussion of the third approach is deferred to Chapter 7 in the context of near real-time applications.

5.5.3.5 Partial least squares

The relationship between sub-block PDCs or sensors can be obtained using techniques such as partial least squares (PLS). At least two trajectory or observation matrices are required as suggested in Figures 5.9 and 5.10.

To illustrate this notion, let $\mathbf{X}_f$ and $\mathbf{Y}_f$, of dimensions $m_x \times N$ and $m_y \times N$, respectively be observational data matrices associated with two PDCs. It is assumed that matrices $\mathbf{X}_f$ and $\mathbf{Y}_f$ are statistically pre-processed in order to make the variables comparable, that is, both matrices are centered and normalized independently.

Partial least-squares methods project the data matrices onto low-dimensional score matrices **T** and **U**, respectively as [7, 25]

$$\hat{\mathbf{X}}_f = \mathbf{T}\mathbf{P}^T + \mathbf{E}_x \tag{5.12}$$

and

$$\mathbf{Y}_f = \mathbf{U}\mathbf{Q}^T + \mathbf{E}_y \tag{5.13}$$

where **T** is the $p \times m_x$ matrix of scores, and **P** is the $p \times N$ matrix of loadings associated with the data matrix $\mathbf{X}_f$, and **U** is the $p \times m_y$ matrix of scores, and **Q** is the $p \times N$ matrix of loadings, associated with the data matrix $\mathbf{Y}_f$. Matrices $\mathbf{E}_x$ and $\mathbf{E}_y$ contain residuals associated with the unexplained variance (error terms) in $\mathbf{X}_f$ and $\mathbf{Y}_f$.

It can be shown that the score vectors are related by the linear model

$$\mathbf{U} = \mathbf{B}\mathbf{T} + \mathbf{R}_u \tag{5.14}$$

where **B** is the m-by-P matrix of scores from the PLS decomposition.

Several observations are of interest here:

- The $m_y \times m_y$ matrix of coefficients **B** indicates which measurement regions are important for prediction and correlation.
- The residuals provide information about outlier detection and can be estimated as

$$\mathbf{E}_x = \hat{\mathbf{X}}_f - \mathbf{T}\mathbf{P}^T \tag{5.15}$$

and

$$\mathbf{E}_y = \mathbf{Y}_f - \mathbf{U}\mathbf{Q}^T \tag{5.16}$$

A useful additional step consists of post-processing the individual models using SVD analysis.

Other measures of interest to power system data are being developed. The reader is referred to [25] for details about this approach.

5.5.4 Nonlinear PCA

Nonlinear PCA is an extension of PCA analysis which is used to construct a low-dimensional space, which represents the nonlinear characteristics of a high-dimensional space [26].

With reference to (5.9), the idea is to replace the linear mapping by a nonlinear vector function. By analogy with (5.9) one seeks a mapping of the form

$$\mathbf{y} = \mathbf{g}(\mathbf{X}) \tag{5.17}$$

where $\mathbf{g}$ is a nonlinear vector function composed of m nonlinear functions $\mathbf{g} = [g_1(\mathbf{x}), g_2(\mathbf{x}), \ldots, g_m(\mathbf{x})]$.

5.5.5 Blind source separation

A natural extension to the single-block PCA model can be obtained by posing the problem of mode identification within the framework of blind source separation (BSS) techniques [27, 28].

The BSS problem refers to the process of extracting unobserved (source) signals from a set of observations without additional information about the individual sources or the mixing process.

More formally, let $\{s_i(t), i = 1, \ldots, n\}$ be a set of n source signals (i.e., the oscillatory modal content that must be estimated), which are assumed to be statistically independent and to have a zero mean.

Let now the vector $\mathbf{x}(t) = [x_1(t)\ x_2(t)\ \cdots\ x_m(t)]^T$ represents a set of snapshots obtained from the observed data at m system locations ($m \geq n$), sampled at time $t = t_k$, $k = 1, \ldots, N$, which are assumed to be represented as a sum of weighted source signals contaminated by background noise [23]:

$$\mathbf{x}(t) = \mathbf{Hs}(t) + v(t) = \begin{bmatrix} h_{11} & h_{12} & \cdots & h_{1n} \\ h_{21} & h_{22} & \cdots & h_{2n} \\ \vdots & \vdots & \ddots & \vdots \\ h_{m1} & h_{m2} & \cdots & h_{mn} \end{bmatrix} \begin{bmatrix} s_1(t) \\ s_2(t) \\ \vdots \\ s_n(t) \end{bmatrix} + \begin{bmatrix} v_1(t) \\ v_2(t) \\ \vdots \\ v_m(t) \end{bmatrix} \tag{5.18}$$

where $\mathbf{H} = [\mathbf{h}_1\ \mathbf{h}_2\ \cdots\ \mathbf{h}_n]$ is an m-by-n unknown mixing matrix representing the stationary linear transformation from the n-dimensional vector of source signals $\mathbf{s}(t) = [s_1(t)\ s_2(t) \cdots s_n(t)]^T$ to the ensemble of observations $\mathbf{x}(t)$, and the m-dimensional vectors of measurement noise $v(t) = [v_1(t)\ v_2(t)\ \cdots\ v_m(t)]^T$.

Matrix $\mathbf{H}$ represents the relationship between the measured responses (snapshots) and the source signals. Physically, matrix $\mathbf{H}$ may be viewed as a transformation matrix between the time domain and the feature data. In words, the kth source signal $s_k(t)$ is directly related to the kth feature in the vibration data.

As discussed below, the columns of the mixing matrix provide a measure of observability while the source signals contain the associated natural frequencies and damping ratios. Evaluating (5.18) at each snapshot, the ensemble of observations $\mathbf{X}(t) = [\mathbf{x}(t_1)\ \mathbf{x}(t_2)\ \cdots\ \mathbf{x}(t_N)]^T \in \mathbb{R}^{m\times N}$ can be written as

$$\mathbf{X}(t) = [\mathbf{x}(t_1)\quad \mathbf{x}(t_2)\quad \cdots\quad \mathbf{x}(t_N)]^T = \mathbf{HS}(t) + \mathbf{Y}(t) \tag{5.19}$$

where $\mathbf{S}(t) = [\mathbf{s}(t_1)\ \mathbf{s}(t_2) \cdots \mathbf{s}(t_N)] \in \mathbb{R}^{n\times N}$ and $\mathbf{Y}(t) = [\upsilon(t_1)\ \upsilon(t_2) \cdots \upsilon(t_N)] \in \mathbb{R}^{m\times N}$. For uniqueness of the decomposition, constraints are applied to both the mixing matrix and the source signal such as sparsity or interdependence of the components.

The source separation problem can then be defined as the simultaneous estimation of the mixing matrix $\mathbf{H}$, and the underlying oscillatory modal components associated with the inter-system oscillations from the observed noisy measurements, using only the measured data vector.

The BSS problem can be formulated as finding a demixing matrix $\mathbf{H}^{-1}$ such that the output $\hat{\mathbf{s}}(t) = \mathbf{H}^{-1}\mathbf{x}(t)$ is the best approximation to the source signals, $\mathbf{s}(t)$, in (5.18). Once the source signals and mixing matrix have been estimated, the original time histories from spatial sensors can be reconstructed. Figure 5.11 provides a schematic representation of this model.

This approach, while straightforward, suffers from a number of disadvantages including a susceptibility to mode-mixing, the generation of spurious components, and the requirements of a preexisting knowledge of the number of delays to be used.

5.5.5.1 Lagged variables

Conventional BSS is best suited for the analysis of steady-state data with uncorrelated measurements [27]. This approach may be extended to modeling and monitoring of dynamic measurements by augmenting the data matrix by including lagged variables. This yields

$$\mathbf{X}(t) = [\mathbf{x}_1(t)\quad \mathbf{x}_1(t-1)\quad \cdots\quad \mathbf{x}_2(t)\quad \mathbf{x}_2(t-1)\quad \cdots]^T$$

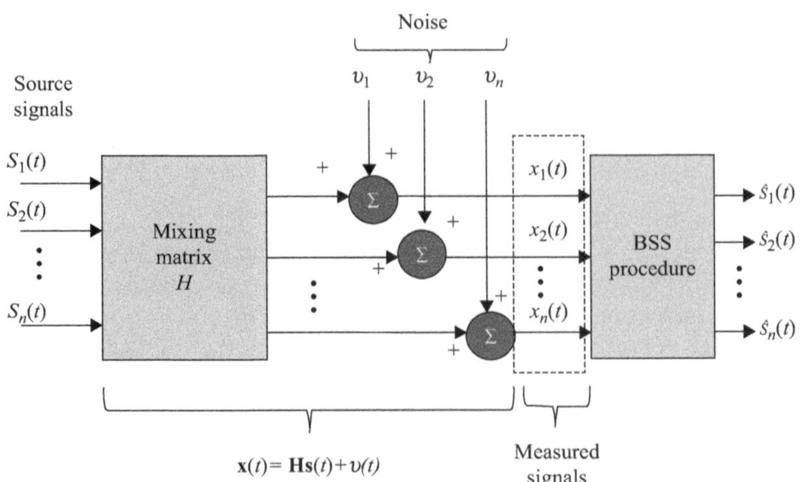

Figure 5.11 A schematic illustration of the BSS method

The method requires prior knowledge or assumptions about the order of the dynamics and increases the computational complexity of the modeling. The reader is referred to Ayón *et al.* [27] for further details on this technique.

A number of practical problems need to be addressed in implementing the above strategy. The first is the selection of the number of delays. Another practical aspect of the implementation of these techniques concerns computational effort.

5.5.5.2 Complex BSS formulations

A second way to handle changing dynamic conditions is based on Hilbert analysis. As pointed out previously, conventional BSS analysis lacks phase information and cannot be used for coherency identification or the analysis of traveling waves. In [29, 30], approaches to incorporate phase information were proposed.

This section extends BSS analysis to the complex case. In order to introduce the general case, consider a scalar field (a sequence of measurements), $x_k(t)$, $k = 1, \ldots, m$. Application of the Fourier transform gives

$$x_k(t) = \sum_{\omega} a_k(\omega)\cos(\omega t) + b_k(\omega)\sin(\omega t) \tag{5.20}$$

for $k = 1, \ldots, m$, where k is the kth grid position, and $a_k(\omega)$ and $b_k(\omega)$ are the Fourier coefficients:

$$a_k(\omega) = \frac{1}{T}\int_0^T x_k(t)\cos(\omega t)dt$$

$$b_k(\omega) = \frac{1}{T}\int_0^T x_k(t)\sin(\omega t)dt$$

Consider now the transformation

$$\hat{x}_k(t) = \sum c_k(\omega)e^{-j\omega t} \tag{5.21}$$

where $c_k = a_k(\omega) + jb_k(\omega)$. Expanding (5.21) yields

$$\begin{aligned}\hat{x}_k(t) &= \textstyle\sum_{\omega}\{[a_k(\omega)\cos(\omega t) + b_k(\omega)\sin(\omega t)] + j[b_k(\omega)\cos(\omega t) - a_k(\omega)\sin(\omega t)]\}\\ &= x_k(t) + j\hat{x}_k(t) = x_k(t) + jH(x_k(t))\end{aligned}$$

where

$$H(x) = \frac{1}{\pi}\int_{-\infty}^{\infty}\frac{x(t)}{t-\tau}d\tau$$

is the Hilbert transform of x and $j = \sqrt{-1}$. This is equivalent to passing the time series x through a filter with a frequency response function (a phase shift of $\pm\pi$) in the spectral domain

$$H(\omega) = \begin{cases} -j, & \text{for } \omega \geq 0 \\ j, & \text{for } \omega \leq 0 \end{cases}$$

Techniques to compute the Hilbert transform are discussed in Chapter 7.

Denote $H(\mathbf{X})$ as the Hilbert transform of the observation matrix $\mathbf{X}$. A complex formulation can be obtained by defining the complex data matrix

$$\hat{\mathbf{X}} = \mathbf{X} + jH(\mathbf{X}) \tag{5.22}$$

where $H(\mathbf{X})$ is the Hilbert transform of $\mathbf{X}$, and is the complex-valued observation matrix.

The complex observation matrix has a number of interesting properties that make it appealing for the analysis of measured data. Let the complex observation matrix be rewritten as $\hat{\mathbf{X}} = \mathbf{X}_R + j\mathbf{X}_I$, where the subscripts R and I refer to the corresponding real and imaginary parts. The covariance matrix is then given by [31]

$$\mathbf{C} = \mathbf{C}_R + j\mathbf{C}_I = \left[\hat{\mathbf{X}}_R^T\hat{\mathbf{X}}_R + \hat{\mathbf{X}}_I^T\hat{\mathbf{X}}_I\right] + j\left[\hat{\mathbf{X}}_R^T\hat{\mathbf{X}}_I - \hat{\mathbf{X}}_I^H\hat{\mathbf{X}}_R\right] \tag{5.23}$$

It can then be readily seen that $\mathbf{C}_R$ is a symmetrical matrix, and that $\mathbf{C}_I$ is an asymmetric matrix or hemisymmetric matrix. Since the symmetrical matrix is a particular case of the Hermitian matrix, all of its eigenvectors are real; the elements of the asymmetrical matrix are all purely imaginary and its eigenvectors are complex conjugate.

The bases for the complex autocorrelation matrix $\mathbf{C}$, are now defined by as $\varphi = \varphi_R(x) + j\varphi_I(x)$. Because the scalar field is then complex they can be represented by their amplitude and phase.

Thus, the vector field can be expanded in terms of the complex coefficients. As discussed in [31], the resulting matrix is Hermitian and therefore possesses a set of complex eigenvectors. The eigenvectors (mode shapes) are determined from the eigenvalue problem $\mathbf{R}\mathbf{\Phi} = \Lambda\mathbf{\Phi}$, where $\mathbf{R} = \mathbf{X}^T\mathbf{X}$, and

$$\mathbf{\Phi} = [\varphi_1 \quad \varphi_2 \quad \cdots \quad \varphi_m]$$

Alternatively, the SVD decomposition of $\hat{\mathbf{X}}$ can be expressed as

$$\hat{\mathbf{X}} = \mathbf{U}\mathbf{\Sigma}\mathbf{V}^H$$

where the superscript H denotes conjugate transpose, the columns of $\mathbf{U}$ are eigenvectors of $\hat{\mathbf{X}}\hat{\mathbf{X}}^H$, and the columns of $\mathbf{V}$ are eigenvectors of $\hat{\mathbf{X}}^H\hat{\mathbf{X}}$.

Analytical development in Chapter 3 has shown that the time evolution of the observation matrix, $\hat{\mathbf{X}}$, can be expressed as

$$\hat{\mathbf{X}} = \mathbf{\Phi}\mathbf{A}(t)$$

with

$$\begin{cases} \mathbf{\Phi} = \mathbf{U} \\ \mathbf{A}(t) = \mathbf{\Sigma}\mathbf{V}^H \end{cases}$$

where $\mathbf{A}(t)$ and $\mathbf{\Phi}$ are complex valued.

In analogy with the conventional POD analysis, the time evolution of the *j*th sensor can be reconstructed as

$$x_j(t) = \mathrm{Re}\left[\sum_{i=1}^{p} \hat{a}_i(t)\hat{\varphi}_i(x)\right] = \mathrm{Re}\left[\sum_{i=1}^{p} |\hat{a}_i|\ |\hat{\varphi}_i(x)| e^{j\left[\varphi_{a_i}(t)+\theta_{\varphi_i}(x)\right]}\right] \tag{5.24}$$

for $j=1, \ldots, N$, where $\hat{a}_i(t)$ and $\hat{\varphi}_i(\mathbf{x})$ are the time-dependent orthogonal time coefficients and the $\varphi_{a_i}(t) + \varphi_{a_\varphi}(x)$ are the spatially dependent basis functions or spatial phase functions.

Four measures that define possible moving features in the scalar or vector fields can be defined as follows [32, 33]:

1. *Spatial amplitude function* $S_j(x)$: The spatial amplitude function shows the spatial distribution of variability associated with each eigenmode and is defined by

$$S_j(x) = \sqrt{\hat{\varphi}_i(\mathbf{x})\hat{\varphi}_i^*(\mathbf{x})} \tag{5.25}$$

   or, in terms of the SVD decomposition, $S_j(x)$ can be defined as the real part of the complex matrix **A**.
2. *Spatial phase function* $\theta_i(x)$: This function shows the relative phase fluctuation among the various spatial locations where x is defined and is given by

$$\theta_i(x) = \arctan\left(\frac{\mathrm{Im}\left|\theta_{\varphi_i}(\mathbf{x})\right|}{\mathrm{Re}\left|\theta_{\varphi_i}(\mathbf{x})\right|}\right) \tag{5.26}$$

3. *Temporal amplitude function* $R_i(t)$: This function measures the temporal variability in the magnitude of the modal structure of the field and can be obtained as

$$R_i(t) = \sqrt{a_i(t)\ a_i^*(t)} \tag{5.27}$$

4. *Temporal phase function* $\varphi_i(t)$: This function describes the temporal variation of phase associated with ⌈ (x, t) and is given by

$$\varphi_i(t) = \arctan\left(\frac{\mathrm{Im}|\varphi_i(t)|}{\mathrm{Re}|\varphi_i(t)|}\right) \tag{5.28}$$

Equations (5.25) through (5.28) provide a complete characterization of any propagating features and periodicity in the original data field. Table 5.3 provides details of its computation in the framework of SVD. The complex oscillation patterns (mode shapes) can now be extracted from the spatial phase functions as

$$\varphi_i(x) = \begin{bmatrix} |\varphi_{i_1}| \angle\theta_{\varphi_{i_1}} \\ |\varphi_{i_2}| \angle\theta_{\varphi_{i_2}} \\ \vdots \\ |\varphi_{i_n}| \angle\theta_{\varphi_{i_n}} \end{bmatrix}, \quad i = 1, \ldots, p \tag{5.29}$$

Table 5.3 summarizes the computation of spatio-temporal measures using the SVD framework.

The combined application of these ideas with other recent approaches such as Koopman mode analysis and dynamic mode decomposition is expected to enhance predictive techniques.

5.6 Other approaches to multisensor data fusion

Other approaches to multisensor data fusion include multiensemble Kalman filtering and the multichannel ARMAX model in [33, 34]. These multisensor data fusion procedures are illustrated on measured frequency data obtained from a six-area interconnected system as detailed in section 4.6. Figure 5.12 shows the detrended frequency records obtained using the HHT procedure in Chapter 4. Similar results are obtained using wavelet multiscale denoising.

Two cases are considered:

1. Characterization of modal behavior from synchronized measurements
2. Analysis of phase relationships

The POD method is first used to estimate mode shapes from multiple synchrophasors; the elements of the frequency data matrix to be

$$\mathbf{X} = \begin{bmatrix} \hat{\mathbf{f}}_1 & \hat{\mathbf{f}}_2 & \hat{\mathbf{f}}_3 & \hat{\mathbf{f}}_4 & \hat{\mathbf{f}}_5 & \hat{\mathbf{f}}_6 \end{bmatrix}^T$$

where $\mathbf{f}_k = [f_{PMU_k}(t_1) \;\; f_{PMU_k}(t_2) \;\; \ldots \;\; f_{PMU_k}(t_N)]^T$, $k = 1, \ldots, 18$ is a column vector of frequency measurements.

Table 5.3 Spatio-temporal measures of system activity

Function	Analytical characterization
Spatial amplitude function $S_i(x)$	$S_i(s) = \text{Re}\,[\mathbf{A}] = \text{Re}\,[\mathbf{U\Sigma}]$
Spatial phase function $\theta_i(x)$	$\theta_i(x) = \arctan\left(\frac{\text{Im}\lvert A_i(x)\rvert}{\text{Re}\lvert A_i(x)\rvert}\right)$
Temporal amplitude function $R_i(t)$	$R_i(t) = \text{Re}\,[\mathbf{B}] = \text{Re}\,[\mathbf{V\Sigma}^T]$
Temporal phase function $\varphi_i(t)$	$\varphi_i(t) = \arctan\left(\frac{\text{Im}\lvert B_i(t)\rvert}{\text{Re}\lvert B_i(t)\rvert}\right)$

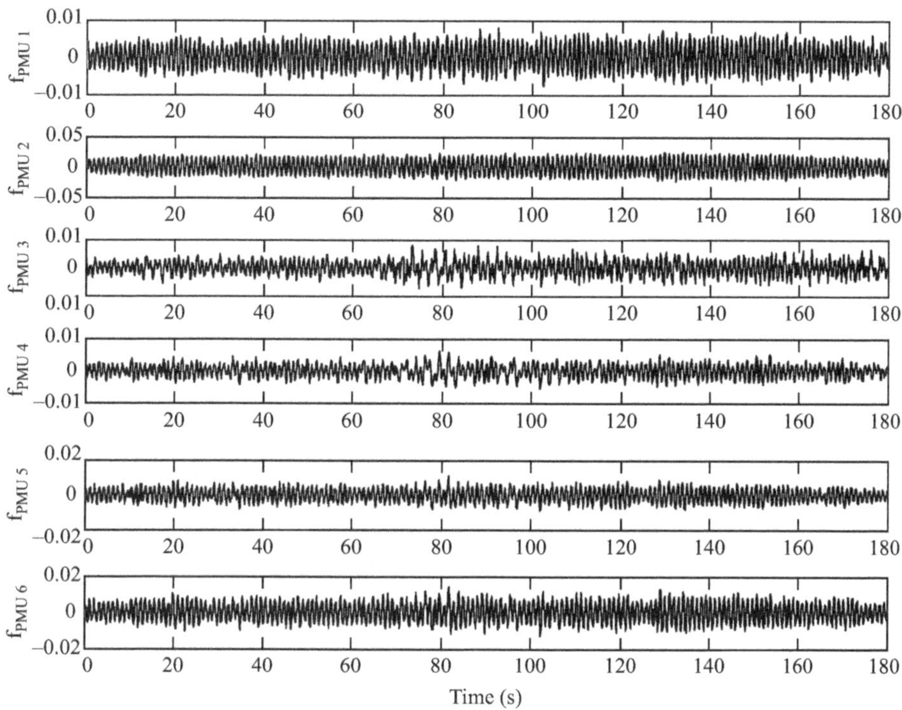

Figure 5.12 Detrended frequency traces. Note the different scales

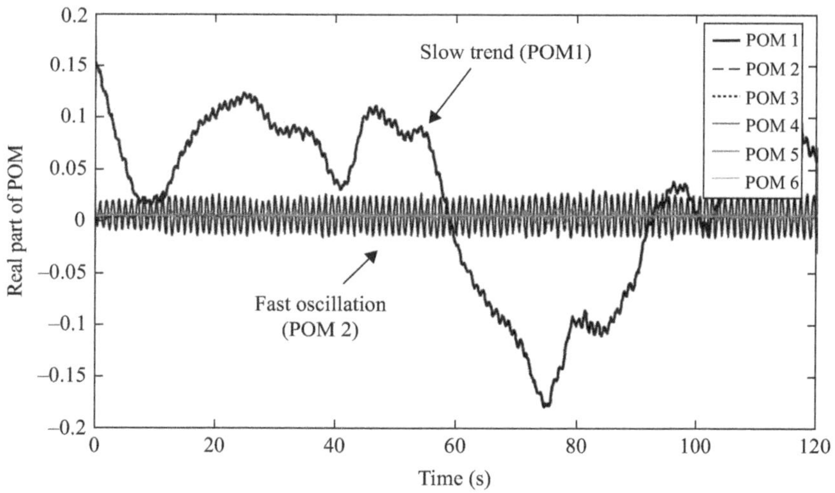

Figure 5.13 Proper orthogonal modes

Figure 5.13 shows the extracted POD modes (POMs), while Figure 5.14 shows the source signals $s(t)$ extracted using the BSS procedure in section 5.5.3. For comparison, Figure 5.15 shows the Koopman modes identified by the procedure in section 4.4.2.

Several observations are of interest here. The first POM captures the slow trend in the signal and is a good approximation to the signal trend. Physically,

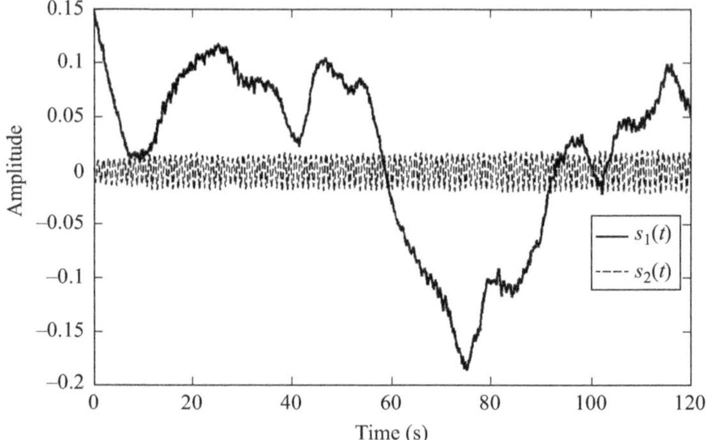

Figure 5.14 Source signals, s(t), extracted using BSS

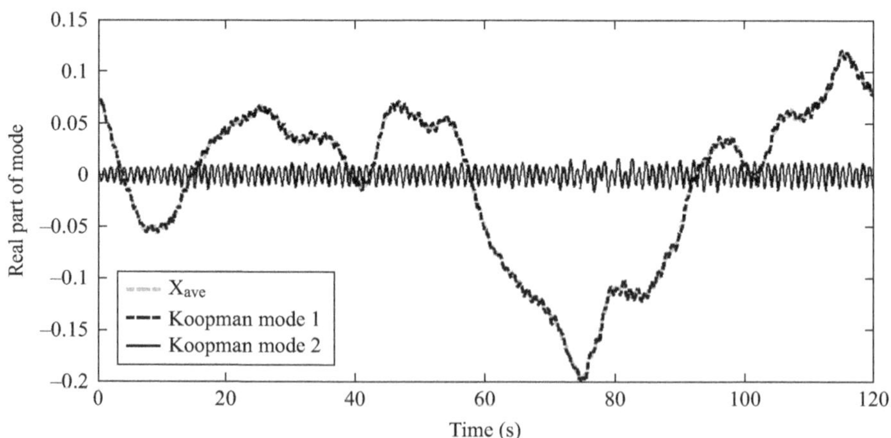

Figure 5.15 Koopman mode decomposition

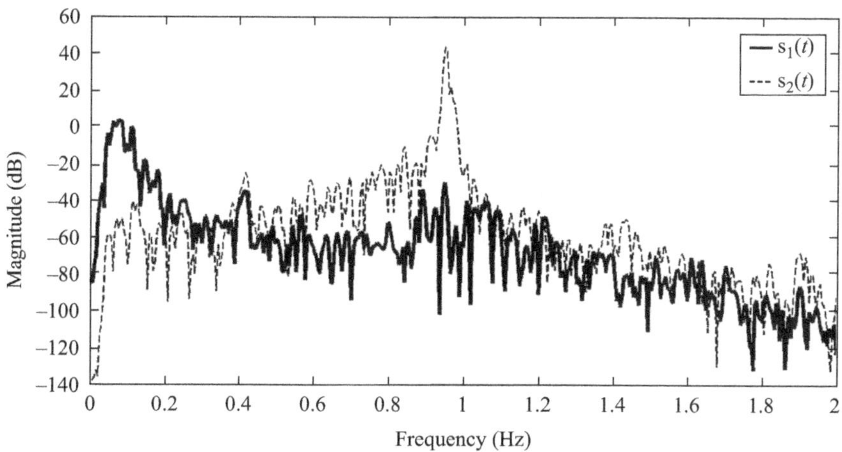

Figure 5.16 Spectra of source signals

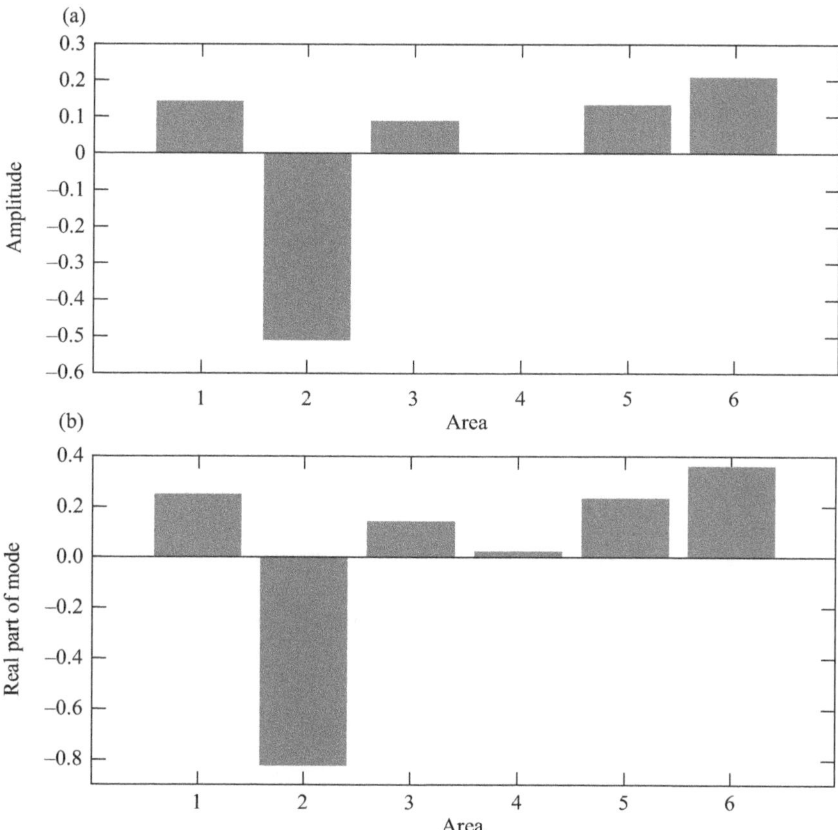

Figure 5.17 Mode shape for the 0.95 Hz mode: (a) BSS; (b) Koopman

the first mode (POM 1) represents the average value of the physical variables associated with measurements, that is

$$a_1(t) = s_1(t) = f_{ave}(t) = \frac{1}{6}\sum_{j=1}^{6} f_j(t) \tag{5.30}$$

In turn, POM 2 is seen to capture the fast variation in the signal in agreement with the fast Fourier transform results in Figure 5.16.

Figure 5.17a shows the corresponding mode shape for the 0.95 Hz mode. For comparison, the mode shape extracted using Koopman mode analysis is presented in Figure 5.17b. Again, these results are found to be consistent with the results obtained using other approaches in Chapter 4.

References

1. C. Reinsel, *Elements of Mutivariate Time Series Analysis*, Springer Series in Statistics, Springer-Verlag, New York, NY, 1997.
2. David L. Hall, James Llinas, 'An introduction to multisensor data fusion', *Proceedings of the IEEE*, vol. 85, no. 1, January 1997, pp. 6–23.
3. A. Sinha, H. Chen, D. G. Danu, T. Kirubarajan, M. Farooq, 'Estimation and decision fusion: A survey', *Neurocomputing*, vol. 71, no. 13–15, August 2008, pp. 2650–2656.
4. K. Worden, W. J. Staszewwski, J. J. Hensman, 'Natural computing for mechanical systems research: A tutorial overview', *Mechanical Systems and Signal Processing*, vol. 25, no. 1, 2011, pp. 4–111.
5. Arturo R. Messina (ed.), *Inter-area Oscillations in Power Systems – A Nonlinear and Nonstationary Perspective*, Power Electronics and Power Systems Series, Springer Science, New York, NY, 2009.
6. Arturo R. Messina, Noé Reyes, Ismael Moreno, Marco A. Perez G., 'A statistical data-fusion-based framework for wide-area oscillation monitoring', *Electric Power Components and Systems*, vol. 42, nos. 3–4, 2014, pp. 396–407.
7. Junshan Zhang, Vijay Vittal, Peter Sauer, 'Networked information gathering and fusion of PMU data – Future grid initiative white paper', Power Systems Engineering Research Center, PSERC Publication 12-07, May 2012.
8. Keith Worden, Wieslaw J. Staszewski, James L. Hensman, 'Neural computing for mechanical systems research: A tutorial overview', *Mechanical Systems and Signal Processing*, vol. 25, 2011, pp. 4–111.
9. David L. Hall, James Llinas (eds.), *Handbook of Multisensor Data Fusion*, CRC Press, Boca Raton, FL, 2001.
10. Bahador Khaleghi, Alaa Khamis, Fakreddine O. Karrah, Saideh N. Razavi, 'Multisensor data fusion: A review of the state-of-the-art', *Information Fusion*, vol. 14, 2013, pp. 28–44.

11. K. Worden, J. M. Dulien Barton, 'An overview of intelligent fault detection in systems and structures', *Structural Health Monitoring*, vol. 3, no. 1, 2004, pp. 85–98.
12. Alex Pappachen Chen, Belur V. Dasarathy, 'Medical image fusion: A survey of the state of the art', *Information Fusion*, vol. 19, 2014, pp. 4–19.
13. Tapio Schneider, 'Analysis of incomplete climate data: Estimation of mean values and covariance matrices and imputation of missing values', *Journal of Climate*, vol. 14, 2001, pp. 853–871.
14. Daniel J. Trudnowski, John W. Pierre, Ning Zhou, John F. Hauer, Manu Parashar, 'Performance of three mode-meter block-processing algorithms for automated dynamic stability assessment', *IEEE Transactions on Power Systems*, vol. 23, no. 2, May 2008, pp. 680–690.
15. G. Ledwich, D. Geddey, P. O'Shea, 'Phasor measurement units for system diagnosis and load identification in Australia', 2008 IEEE Power and Energy Society General Meeting.
16. Paul Nomikos, John F. MacGregor, 'Monitoring batch processes using multiway principal component analysis', *AIChE Journal*, vol. 40, no. 8, August 1994, pp. 1361–1375.
17. Bhavik R. Bakshi, 'Multiscale PCA with application to multivariate statistical process monitoring', *AIChE Journal*, vol. 47, no. 7, 2004, pp. 1596–1610.
18. Manish Misra, H. Henry Yue, S. Joe Qin, Cheng Ling, 'Multivariate process monitoring and fault diagnosis by multi-scale PCA', *Control Engineering Practice*, vol. 26, 2002, pp. 1281–1293.
19. M. A. Kramer, 'Auto-associative neural networks', *Computer Chemical Engineering*, vol. 16, no. 4, 1992, pp. 502–517.
20. M. A. Kramer, 'Nonlinear principal component analysis using auto-associative neural networks', *AIChE Journal*, vol. 37, no. 4, 1991, pp. 313–328.
21. Richard J. Bolton, David J. Hand, Andrew R. Webb, 'Projection techniques for nonlinear principal component analysis', *Statistics and Computing*, vol. 13, 2003, pp. 267–276.
22. Messina HHT-PCA, Coastal Engineering 1998, Copenhagen, pp. 1364–1377.
23. C. Liu, 'Gabor-based kernel PCA with fractional power polynomial models for face recognition', *IEEE Transactions on Pattern Analysis and Machine Intelligence.*, vo. 26, no. 5, May 2004, pp. 572–581
24. Noriaki Hashimoto, Toshihiko Nagai, Masanobu Kudaka, 'Statistical wave forecasting through Kalman filtering combined with principal component analysis', Coastal Engineering 1998, Copenhagen, pp. 1364–1377.
25. S. J. Qin, T. J. McAvoy, 'Nonlinear PLS modeling using neural networks', *Computers in Chemical Engineering*, vol. 16, no. 4, 1992, pp. 379–391.
26. Luis B. Almeida, *Nonlinear Source Separation*, Morgan & Claypool Publishers, San Rafael, CA, 2006.
27. J. J. Ayón, E. Barocio, A. R. Messina, 'Blind extraction and characterization of power system oscillatory modes', *Electric Power Systems Research*, vol. 119, 2015, pp. 54–65.

28. A. Belouchrani, A. Cichocki, 'Robust whitening procedure in blind source separation context', *Electronics Letters*, vol. 24, November 2000, pp. 2050, 2051.
29. A. R. Messina, V. Vittal, 'Extraction of Dynamic Patterns from Wide-Area Measurements using Empirical Orthogonal functions', *IEEE Transactions on Power Systems*, vol. 22, no. 2, May 2007, p. 682.
30. T. P. Barnett, 'Interaction of the monsoon and the pacific trade wind system at interannual time scales. Part I: The equatorial zone', *Monthly Weather Review*, vol. 111, April 1983, pp. 756–773.
31. A. R. Messina, P. Esquivel, F. Lezama, 'Time-dependent statistical analysis of wide-area time-synchronized data', *Mathematical Problems in Engineering*, vol. 2010, 2010, pp. 1–13.
32. M. A. Merrifield, R. T. Guza, 'Detecting propagating signals with complex empirical orthogonal functions: A cautionary note', *Journal of Geophysical Oceanography*, vol. 20, 1990, pp. 1628–1633.
33. Luke Dosiek, John W. Pierre, 'Estimating electromechanical modes and mode shapes using the multichannel ARMAX model', *IEEE Transactions on Power Systems*, vol. 28, no. 2, May 2013, pp. 1950–1959.
34. C. Gao, H. Wang, E. Weng, S. Lakshmivarahan, Y. Zhang, Y. Luo, 'Assimilation of multiple data sets with the ensemble Kalman filter to improve forecasts of forest carbon dynamics', *Ecological Applications*, vol. 21, no. 5, July 2011, pp. 1461–1473.

Chapter 6

Monitoring the status of the system

6.1 Introduction

Accurate diagnosis of system health is a vital step in wide-area monitoring. Advanced event characterization is crucial for improving detection, identification, and characterization of system health. Large interconnected power systems and the systems within them are highly complex and variable structures that defy predictions. Monitoring these systems in the face of uncertainty and variability remains a daunting challenge.

The last two decades have borne witness to an explosion of interest in the development of power system monitoring and analysis techniques [1]. By monitoring the time evolution of key system parameters, monitoring techniques can be used to trigger remedial control actions and alarms, and to aid in the development of situational awareness tools [1–3].

Central to this framework are the diagnostic and prognostic signal processing and measurement techniques used to detect and diagnose power system health [4, 5]. Inappropriate monitoring strategies can lead to irrelevant or poor system characterization, which, in turn, can have profound operational and economic impacts.

Power system monitoring encompasses a variety of activities that involve event detection and classification, and assessment of power system health status [6]. The inclusion of spatio-temporal dynamics is needed in order to identify localized and propagating features in measured data as well as to compress system information. It has been realized that these measurements may contain moving patterns, and travelling waves of different spatial scales and temporal frequencies [7].

Further, because wide-area measurements are characterized by nonlinearity and high dimensionality, a challenging task is to find ways to reduce system dimensionality to a few modes and to link these modes to the underlying dynamical/physical behavior involved.

In this chapter and in Chapter 7, several tools to assess power system health are developed and tested. Methods to assess changes in measured oscillatory response are examined and new approaches for use in wide-area system monitoring are presented.

Issues related to robustness of the methods in the presence of measurement noise and multiple events are discussed.

6.2 Power system health monitoring

In recent years, different threat monitoring techniques have been developed to detect abnormal operation and assess system health including trigger algorithms and blackhole monitoring. Given a sequence of observed data collected from one or more sensors, the problem of power system health monitoring involves the solution of three distinct problems: event detection and location, and assessing the magnitude and extent of system degradation.

Power system health monitoring implies a network of sensors that monitor the behavior of the system online. This paradigm can be described as a four-part process: (1) operational evaluation, (2) data acquisition and cleansing, (3) feature extraction and data reduction, and (4) statistical model development.

Intelligent techniques may also be needed to determine the type and severity of the fault in control and protection applications, as well as to compare and discriminate data sets consisting of high-dimensional data.

Conventionally, the definition of damage in many applications implies a comparison between two different states of the system, that is, a healthy state and a damaged state [8]. In near real-time applications involving recorded measurements, the inference process should be based on measured data directly.

The information collected from the monitoring system may be used to estimate key features of interest such as modal signatures and nonstationarities. Modal parameters, notably modal damping and modal frequency, are sensible indicators of wide-area power system health quality and are commonly used in many real-time monitoring systems [9].

As data sets on spatio-temporal processes grow increasingly larger, methods for their statistical analysis within a realistic time frame become tremendously important. The use of adaptive, data-driven monitoring techniques is expected to add important information to current data fusion strategies which can be of interest in the monitoring and control of transient processes in large interconnected power systems. In particular, these methods may be used to monitor the health status and instability risks and for the early detection, isolation, and diagnosis of system threats. These frameworks involve four major steps:

1. Creation of real-time spatio-temporal databases
2. Disturbance detection and characterization
3. Feature extraction, selection, and classification
4. Self-diagnostic and prognosis capability to distinguish between health and fault conditions

Steps 2-4 are discussed separately in the following sections.

6.3 Disturbance and anomaly detection

In devising an intelligent fault event and anomaly detection system, a primary consideration is a clear identification of when a power system transient has occurred [2].

In power system applications, a dynamic event can be thought of as an instance in time when a significant (and persistent) change in the measured response occurs. In contrast, an outlier or nontypical data may be seen as short yet significant deviation from normal behavior.

Data processing and event detection logics may include the following:

- Detection of abrupt changes
- Detection and classification of the start of a disturbance evolution
- Initiation or logging of event recording
- Tracking and assessment of dynamic trends
- Pattern recognition
- Generation of operator alerts and cross-triggers to other recording facilities

To automatically initiate oscillatory monitoring, a change or event detection strategy is usually adopted and decision rules are incorporated; abnormal or non-typical operation is detected if the measurements deviate from the region of normal operation as discussed below.

Figure 6.1 illustrates graphically this notion [10, 11]. As suggested in this plot, a dynamic event is detected if the amplitude of the measured waveform is above a critical threshold, and the oscillatory response persists for a given time. The choice of the threshold is usually arbitrary and is difficult to select a single threshold value suitable for all contingency scenarios. In practice, thresholds are pre-selected based on operating experience or practical criteria.

A number of well-defined parameters can be used in order to identify damage or situational awareness. These parameters include the following [10]:

1. Maximum amplitude of the signal
2. Start time and duration of the oscillatory process
3. Threshold crossings
4. Rise time and decay time
5. Envelope of the observed oscillation

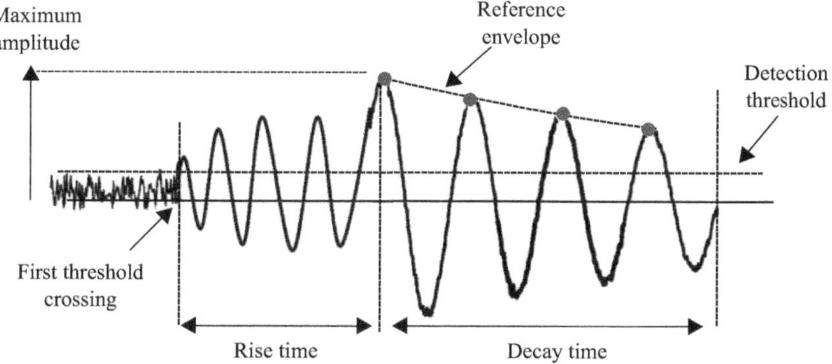

Figure 6.1 Basic parameters for disturbance detection

Several trigger algorithms are based on variations of these ideas [12, 13]. A better strategy is to dynamically select the threshold value to identify the start of a disturbance or anomalous event. A second related problem is to detect the end of the abnormal activity. False alarms, however, should be minimized.

In what follows, techniques to identify and classify damage (location, type, and severity) for a dynamical system exposed to varying environmental and operational conditions as well as instrumentation noise while eliminating false indications are reviewed. These techniques can be classified into the following major categories: (a) modal-based methods, (b) local diagnostic methods, (c) nonparametric methods, (d) time series, nonstationary methods, and (e) statistical pattern recognition methods.

6.4 Modal-based health monitoring methods

Modal-based methods have been widely used to monitor system behavior. Modal frequencies and damping, in particular, are two sensitive indicators of power system stress [14]. In addition, modal phase, modal amplitude, and the associated modal energy can be used to characterize localized phenomena in time and detect system damage.

The use of mode shapes may also be relevant [14] and may signal changes in system behavior, for instance, associated with changes in system topology [14, 15]. An example of this behavior is discussed in [15] in the context of actual measured data.

6.4.1 Filtering and data conditioning

Data pre-processing is usually the first element of signal processing (low-level fusion) for damage detection. It involves normalization, trend removal techniques, detection of change points, outlier analysis, averaging smoothing and filtering, and decimation.

Several complementary approaches to data processing for damage identification can be devised:

1. Damage identification from trending in system behavior
2. Damage identification from changes to model structure
3. Pattern recognition-based approaches to damage identification
4. Entropy-based damage detection methods

In practice, signals are filtered with the objective of extracting the correct set of frequency components that are relevant to this comparison [16, 17]. Filtering is normally performed using linear filters. The objective is to extract noise as well as other artifacts that may affect mode estimation. The analysis of this issue is postponed until Chapter 7 in the context of real-time applications.

An interesting alternative to filtering and denoising can be obtained for signals with multiscale features from the structure of time-series representations in Chapter 4. Consider a sequence of measured data points $x(t_k)$, $k = 0, 1, \ldots, N$.

Multiscale features can be analyzed by expressing the data sequence as a sum of basis functions [18]

$$x(t) = \underbrace{\sum_{j=1}^{p} c_j(t)}_{\text{Noise+HFC}} + \underbrace{\sum_{k=p+1}^{r} c_k(t)}_{\text{Physically meaningful components}} + \underbrace{\sum_{l=r+1}^{n} c_l(t)}_{\text{Artificial components}} \tag{6.1}$$

where the first term represents noise and high-frequency components (HFC). The $c_k(t)$, $k=p+1,\ldots,r$ is the kth modal component associated with the frequency ω_k, and are assumed to be of the form $c_k(t) = A_k(t)\cos(\omega_k t + \varphi_k)$. The last term on the *rhs* of (6.1) represents essentially irregular component trends and unphysical system behavior.

Representations of this type have been obtained in the context of the Hilbert–Huang transform (HHT) and wavelet analysis in previous chapters, but this representation is general and can be obtained from various time-series representations.

It is therefore natural to select the underlying phenomena of interest by discarding insignificant or uninteresting behavior in (6.1) as

$$\hat{x}(t) = x(t) - \sum_{j=1}^{p} c_j(t) - \sum_{l=r+1}^{n} c_l(t) \tag{6.2}$$

where $\hat{x}(t)$ is the bandpass (denoised and detrended) signal, and the index n represents a subset of the modal components obtained by discarding non-important or uninteresting components in (6.2).

In practice, the spurious components c_l, $l=r+1,\ldots$ can be discarded using a suitable threshold or energy criterion. Given estimates of the instantaneous amplitude and frequency, A, and ω at time instant t, several objective criteria, *Trsh*, to measure the contribution of each intrinsic mode function (IMF) to the total energy can be obtained.

Table 6.1 summarizes some commonly used approaches described in recent literature. A more detailed discussion of these issues is provided in subsequent sections.

Ideally, the bandpass signal $\hat{x}(t)$ contains oscillatory components associated with a given frequency band of interest. This intuitive idea has been explored in recent work using linear filters [6].

Three basic applications of this idea can be considered for power system measured data:

1. *Noise reduction:* Subtracting the higher frequency components, c_j, in (6.2), noise can be eliminated or reduced in a systematic manner.
2. *Data adaptive smoothing or filtering:* Selected temporal frequency scales can be removed by subtracting from (6.2) frequency components of concern.
3. *Trend (slowly developing events) extraction:* As a by-product of the proposed procedure, the time-varying mean, $m(t)$, can be systematically extracted and used for global system monitoring.

The outcome of this analysis is a representation of the form

$$\hat{x}(t) = \underbrace{m(t)}_{\substack{\text{Dynamic}\\\text{trend}}} + \underbrace{c(t)}_{\substack{\text{Oscillatory}\\\text{components}}} + \underbrace{\varepsilon(t)}_{\text{Noise}} \tag{6.3}$$

Table 6.1 Measures of signal's strength

Method	Description
Relative amplitude [19]	$Trsh_{IMF_j}(t) = \dfrac{A_j(t)}{\sum\limits_{k=1,k\neq j}^{p} A_k(t)}, \quad j = 1,\ldots,p$
Frequency weighted amplitude [20]	$c_l = \dfrac{\sum\limits_{k=1}^{N} c_l(t_k)\,\omega_j(t_k)}{\sum\limits_{k=1}^{N} c_l(t_k)\,\omega_j(t_k)}$
Entropy [21]	$H = -\lim\limits_{M\to\infty} \dfrac{1}{\log M} \sum\limits_{j=1}^{M} p_j \log p_j$
Norm [22]*	$Trsh_n = \dfrac{\lVert c_j(t)\rVert}{\lVert x(t)\rVert}$

*l-2 norm.

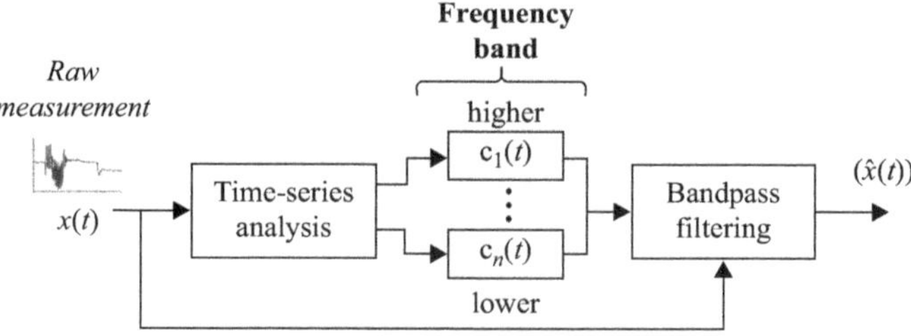

Figure 6.2 Bandpass filtering using time-frequency analysis

in which $m(t)$ represents the time-varying instantaneous mean or underlying trend, $c(t)$ is the fluctuating variation (the oscillatory components) of the signal, and $\varepsilon(t)$ represents noise effects.

The successful implementation of this technique requires the solution of three interrelated problems:

1. Noise reduction
2. The identification and extraction of the true oscillatory components
3. The extraction of the instantaneous mean, $m(t)$

Representations of the form (6.3) are inherent to some time-series representations, that is, DHR and EMD analyses, or can be obtained by appropriate filtering of the data.

Figure 6.2 gives a schematic representation of the analysis for the univariate (single sensor) case motivated by the decomposition procedure discussed in section 4.3.

The algorithmic procedure is formally stated below:

Algorithmic procedure for bandpass filtering

Given a data sequence $x(t_k)$, $k = 1, \ldots, N$.

1. Decompose the signal $x(t_k)$ into a collection of IMFs, $c_1(t), \ldots, c_n(t)$, with associated frequency components $\omega_j(t), j = 1, \ldots, n$.
2. Based on the instantaneous frequencies $\omega_j(t)$, select a frequency band $\omega_{\min} \leq \omega_j(t) \leq \omega_{\max}$.
3. Bandpass the original signal using (6.2).
4. Determine a suitable threshold for automated triggering of alarms and other actions.

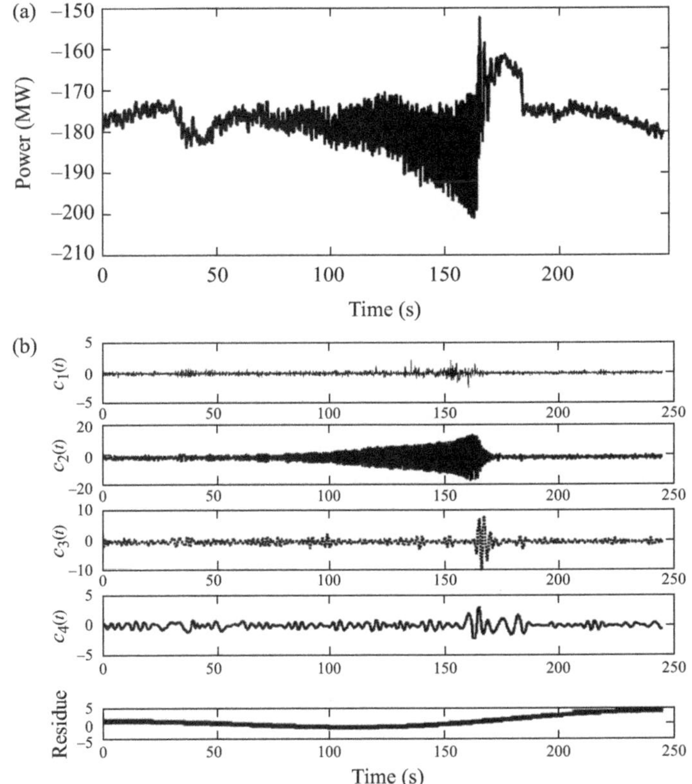

Figure 6.3 Test signal and its modal decomposition: (a) measured signal; (b) modal decomposition

This approach allows for nonstationary behavior to be analyzed into separate frequency bands. This, in turn, facilitates the interpretation of modal behavior in terms of basic modal information.

To illustrate the procedure, consider the measured signal in Figure 6.3a. Application of HHT analysis in Figure 6.3b results in seven IMFs and a trend.

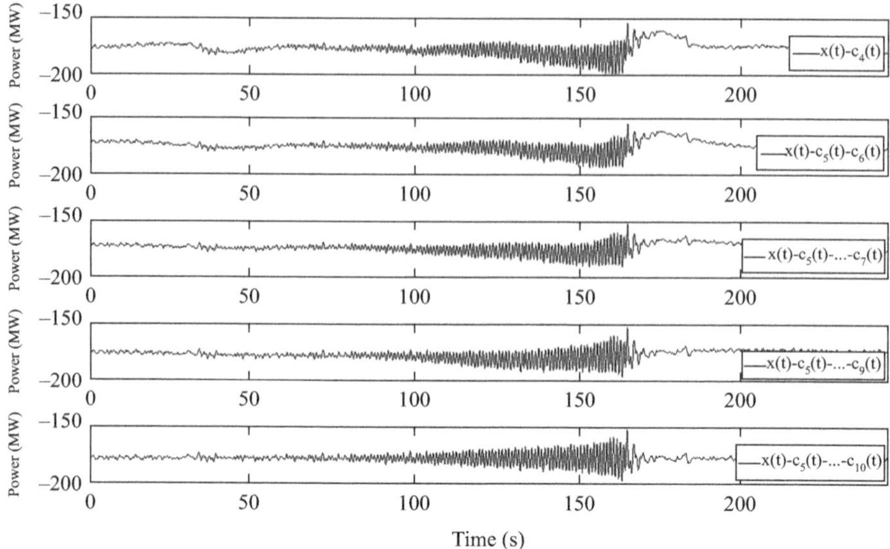

Figure 6.4 Illustration of the detrending (and denoising) process

As discussed in Chapter 5, IMF 1 essentially captures the noise in the signal, while IMF 2 is seen to capture the dominant system behavior. Also of relevance, IMF 3 is seen to capture system behavior following a switching action.

Figure 6.4 demonstrates how the above procedure can be used to detrend or selectively remove unwanted modal components. At each step, of the modal decomposition, unwanted components are removed as

$$\hat{x}_i(t) = x_{i-1}(t) - c_i(t), \quad i = 1, \ldots, n$$

where

$$h_i(t) = \begin{cases} c_i & \text{if } i < i_{\min}, i > i_{\max} \\ 0, & \text{otherwise} \end{cases}$$

and $i_{\min}$, $i_{\max}$ are the indices associated with the frequency band of interest.

In the more general and interesting case, denoising can be performed simultaneously by removing component $c_1(t)$ in a selective manner. The method is particularly well suited to the analysis of multimodal data.

6.4.2 Entropy and energy

Energy and entropy concepts are powerful tools with which to identify changes in system behavior. Following the nomenclature in Chapter 3, consider a sequence of measured data $\mathbf{x}(t) = [x(t_1), x(t_2), \ldots, x(t_N)]^T$. Using time-series theory, the time evolution of the signal can be expressed as

$$\mathbf{x} = \sum_{j=1}^{M} \lambda_j a_j(t) \varphi_j = \sum_{j=1}^{M} \hat{a}_j(t) \varphi_j$$

where the a_j are the temporal coefficients and the φ_j are spatial vectors.

The global mode entropy, H, is defined by [18]

$$H = - \lim_{M \to \infty} \frac{1}{\log M} \sum_{j=1}^{M} p_j \log p_j \tag{6.4}$$

where p_j is the normalized energy probability distribution of the modal component defined as [21]

$$p_j = \frac{\lambda_j^2}{\sum_{j=1}^{M} \lambda_j^2}$$

With these definitions in hand, the corresponding temporal and spatial entropies are defined by

$$H(x) = - \frac{1}{\log M} \sum_{j=1}^{M} p_j(x) \log p_j(x) \tag{6.5}$$

and

$$H(t) = - \frac{1}{\log M} \sum_{j=1}^{M} p_j(t) \log p_j(t) \tag{6.6}$$

respectively, where

$$p_j(x) = \frac{\lambda_j |\varphi_j(x)|}{\sum_{j=1}^{M} \lambda_j |\varphi_j(x)|}; \quad p_j(t) = \frac{\lambda_j |a_j(t)|}{\sum_{j=1}^{M} \lambda_j |a_j(t)|}$$

Two properties of interest can be derived here:

1. The global entropy, H, depends on the number of nonzero eigenvalues.
2. The entropies range from 0 to 1; the entropy is maximal ($H = 1$) if all eigenvalues are equal.

The above formulation extends readily to other time–frequency representations. For instance, the total kinetic energy captured by the jth mode using the HHT and wavelet analysis can be obtained using the procedure described below [23]–[25].

Let, to this end, the measured signal be represented by

$$x(t) = \sum_{j=1}^{n} c_j(t) = c_1(t) + \cdots + c_j(t) + \cdots + c_n(t)$$

The total energy of signal $x(t)$ can be calculated as

$$E(t) = \sum_{j=1}^{n} E_j(t) = \int_{-\infty}^{\infty} x^2(t)dt \approx \int_0^t (IMF_j)^2 dt = \int_0^t (c_j)^2 dt \tag{6.7}$$

where

$$E_j = \int_0^t (IMF_j)^2 dt = \int_0^t (c_j)^2 dt \tag{6.8}$$

and use has been made of the orthogonality properties

$$\int_{-\infty}^{\infty} x_i(t)x_j(t)dt \approx 0, \quad \text{for } i \neq j$$

In a similar manner, the wavelet energy is given by

$$E_j^i = \int_0^t (f_j^i)^2 dt$$

or

$$E_j^i = \int_0^t (f_j^i)^2 dt = \sum_{i=1}^{2J} E_j^i \tag{6.9}$$

for the case of an orthogonal mother wavelet.

Further, the energy fraction of the jth mode is given by

$$\hat{E}_j(t) = \frac{c_j^2(t)}{\sum_{j=1}^{n} c_j^2(t)} = \frac{E_j(t)}{E_T(t)} \tag{6.10}$$

where $E_T = \sum_{j=1}^{n} E_j(t) = \sum_{j=1}^{n} c_j^2(t)$ is the total energy. Similar interpretations can be obtained for POD or other decompositions.

These measures can be used for damage detection, detection of change points, and system monitoring as discussed below. Figure 6.5 compares the modal energies for the temporal components $c_2(t)$ and $c_3(t)$ shown in Figure 6.3 with the total signal entropy calculated from the energy ratio in (6.6). As shown in this plot both modal entropies and energies can be used to identify localized events in time at a given scale.

Also of interest, Figure 6.6 shows the wavelet global spectrum of the measured signal, $x(t)$. These results correlate well with the energy results presented in Figure 6.5, suggesting the potential use of these techniques for mode visualization.

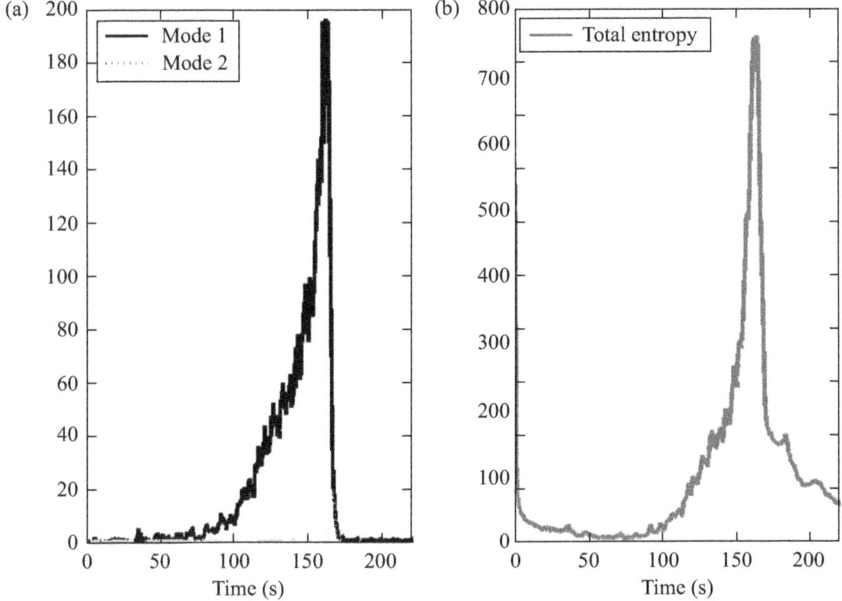

Figure 6.5 Instantaneous modal energy and entropy for the signal in Figure 6.3. Values are not normalized. (a) Modal energy; (b) Entropy

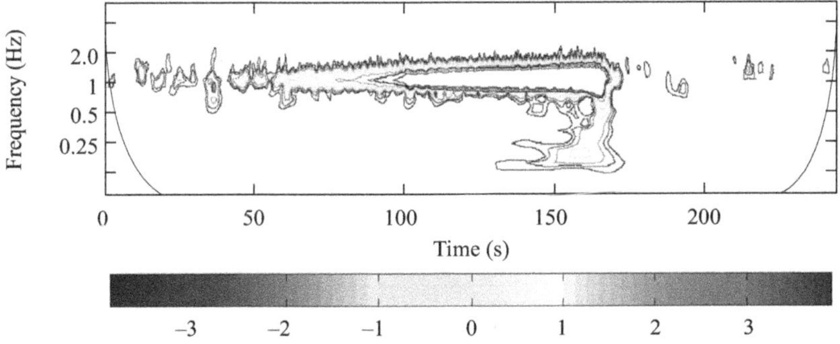

Figure 6.6 Wavelet global spectrum for mode 2 in Figure 6.3

6.4.3 Entropy-based detection of system changes

Extension of the above approach to identify and characterize multiple events is immediate. As a simple example of the application of the above ideas, consider the measured frequency signal shown in Figure 6.7.

The signal represents the time evolution of bus frequency at a major (500 kV) transmission bus following two switching events and is thought to be representative of other events associated with multiple contingencies.

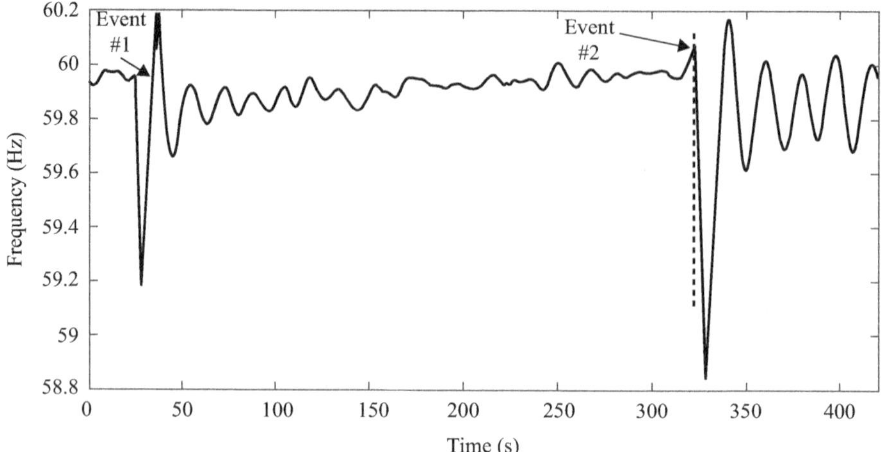

Figure 6.7 Time traces of recorded bus frequency swing showing multiple disturbances followed by transient oscillations. The arrows mark two generation tripping events

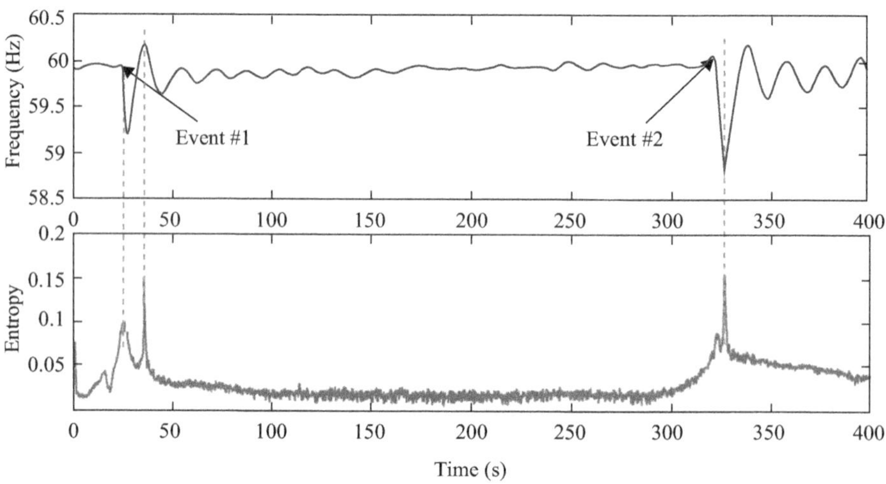

Figure 6.8 Change of total entropy as a function of time for the bus frequency swings in Figure 6.7

Following a similar approach to that of the previous event, the signal is decomposed into a series of mono-component signals as $x(t) = \sum_{j=1}^{n} c_j(t)$. As shown in Figure 6.8, the application of entropy criteria allows to identify abrupt changes in system behavior originating from system disturbances.

The analysis suggests that entropy can be used to decompose a nonstationary signal into stationary or quasi-stationary intervals for adaptive segmentation of measured data. Other, more specific applications of these techniques are presented in [25].

Let us now turn to the problem of assessing global system behavior.

6.5 Wide-area inter-area oscillation monitoring

This section outlines and compares the use of multisignal modal analysis techniques to extract modal properties: multisignal Prony analysis and Koopman mode analysis. Refer to section 4.4 of this book and [26–30] for further details about these methods.

As a simple motivational example, the 5-machine test system used in Chapter 3 is adopted to assess the ability of these techniques to characterize modal behavior. For the purposes of this analysis, ten bus voltage magnitudes (refer to Figure 6.9) are used to extract the main modes of oscillation.

Visual inspection of the time evolution of bus voltage magnitudes in Figure 6.10 shows three clusters of buses exhibiting a nearly common behavior in a sense to be defined more precisely. These are as follows:

Cluster 1 composed of bus 5
Cluster 2 composed of buses 1, 3, 6, 11, and 12
Cluster 3 composed of buses 2, 4, 6, and 7

Two cases are considered:

Case A: Base case with five voltage measurements associated with transmission buses. This is the case studied in Chapter 3.
Case B: Base case with ten voltage measurements.

6.5.1 Case A

Table 6.2 shows modal estimates for the five-voltage measurement case (Refer to Figure 6.10a). For comparison multisignal Prony modal results are also displayed.

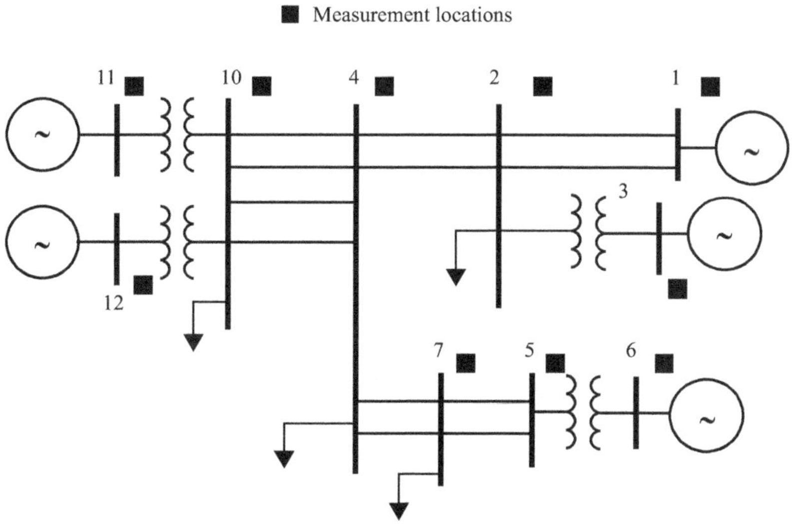

Figure 6.9 Five-machine, ten-bus test system. Filled circles indicate measurement locations

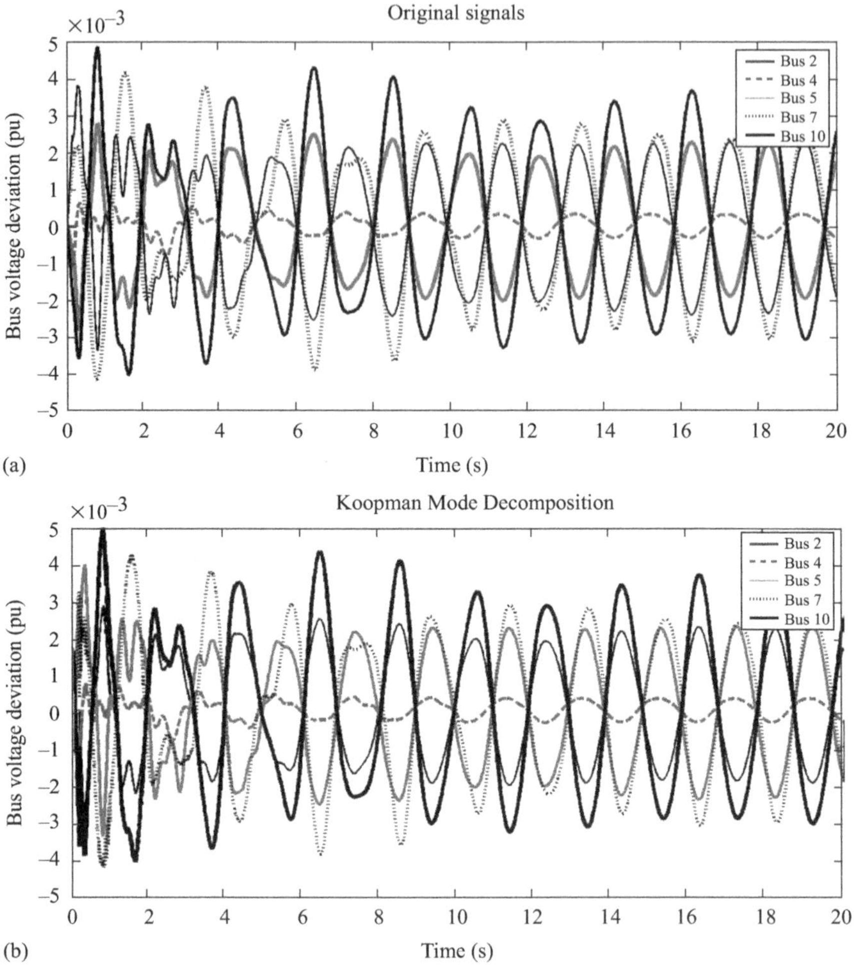

Figure 6.10 Koopman mode approximation: (a) original signals; (b) Koopman mode reconstruction of measured bus voltage signals

Table 6.2 Comparison of modal estimates for voltage signals: Case A

	SSSA*		**MSPA****		**KMA*****	
Mode	f **(Hz)**	$\xi/2\pi$	f **(Hz)**	$\xi/2\pi$	f **(Hz)**	$\xi/2\pi$
1	0.510	−0.0008	0.526	0.02300	0.516	0.0037
2	0.906	0.0246	0.917	0.02340	0.928	0.0310
3	1.497	0.0284	1.498	0.04430	1.456	0.0204

*Small Signal Stability Analysis (SSSA) results
**Multisignal Prony analysis (MSPA) results
***Koopman mode analysis (KMA) results

As shown in Table 6.2, Koopman mode analysis provides a good approximation to system behavior. Discrepancies are noted, especially for the unstable mode at 0.51 Hz. One implication suggested by this analysis is that Koopman mode analysis may not provide a proper characterization of system behavior under incomplete observability of the system.

In addition Koopman mode estimates can only capture average system behavior. More comprehensive simulation and theoretical analyses need to be completed, however, to verify this generalization.

Figure 6.10b compares the original (detrended voltage deviations) and the Koopman mode reconstructions. The maximum error is 4.94910^{-3}.

6.5.2 Case B

In this analysis, the numerically obtained voltage time series at buses 1, 2, 3, 4, 5, 6, 7, 10, 11, and 12 are used to form the observation matrix, which is defined as $\mathbf{X} = [\mathbf{V}_1(t)\ \mathbf{V}_2(t)\ \mathbf{V}_3(t)\ \mathbf{V}_4(t)\ \mathbf{V}_5(t)\ \mathbf{V}_6(t)\ \mathbf{V}_7(t)\ \mathbf{V}_{10}(t)\ \mathbf{V}_{11}(t)\ \mathbf{V}_{12}(t)]^T$, where $\mathbf{V}_j(t)$, $j = 1, \ldots, 10$ is a time vector of bus voltage deviations defined as $\mathbf{V}_j(t) = [V_j(t_1)\ V_j(t_2) \ldots V_j(t_N)]^T$, with $N = 296$ samples.

Figure 6.11 shows the time evolution of the simulated bus voltage magnitudes following the above disturbance.

Table 6.3 compares modal estimates for three different modal approximations:

1. Small signal stability analysis (SSSA) of a linearized system model
2. Multisignal Prony analysis (MSPA) results based on a Kumaresan–Tufts implementation of the model
3. Koopman mode analysis (KMA)

For cases 2 and 3 above, the ten bus voltage magnitude signals were analyzed simultaneously.

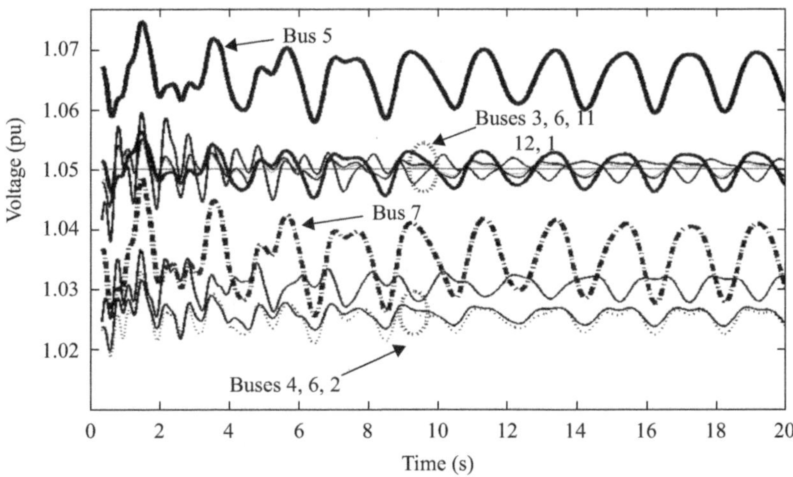

Figure 6.11 Time traces of bus voltage magnitudes

Table 6.3 Comparison of modal estimates for voltage signals: Case B

	SSSA		MSPA		KMA	
Mode	f **(Hz)**	$\xi/2\pi$	f **(Hz)**	$\xi/2\pi$	f **(Hz)**	$\xi/2\pi$
1	0.510	−0.00080	0.510	−0.00048	0.510	−0.00031
2	0.906	0.02460	0.917	0.02340	0.928	0.02450
3	1.497	0.02840	1.498	0.04430	1.499	0.02440

Results in Table 6.3 are found to be highly consistent with main differences in the damping estimates, especially for mode 3. In all cases, estimation errors are below 1% showing that global analysis techniques can be used to extract modal behavior from multiple recording devices.

Comparison of these methods in the context of more complex system oscillations is provided in [30].

For completeness, the set of measurements were analyzed using the diffusion map algorithm and the results are compared to those of Koopman and Prony analyses.

6.6 High-dimensional pattern recognition-based monitoring

Disturbance or anomaly detection in measured data can be regarded as a problem of pattern recognition [6, 31]. In these approaches, patterns represent different classes or patterns of behavior representing disturbance or anomalous conditions.

This section poses the problem of modal extraction of measured dynamic trajectories in the context of a statistical pattern recognition paradigm and explores techniques to analyze, identify, and cluster multimodal data.

Given a measurement matrix **X**, the analysis procedure divides into four principal phases (1): calculation of time-dependent measurement distances, (2) data diffusion, (3) computation of diffusion distances, and (4) extraction of time coordinates. Extensions to the near real-time setting are discussed in Chapter 7.

6.6.1 Sparse diffusion implementation

As mentioned earlier in Chapter 3, the Markov matrix **M** in the diffusion process is numerically full. The feature space, however, is usually sparse. To circumvent these limitations, values below a given threshold can be disregarded and thus leading to a sparse representation.

In developing the algorithm, a singular value decomposition (SVD)-based implementation of the algorithm in Chapter 3 has been adopted. The procedure is briefly summarized below.

Algorithmic procedure for high-dimensional pattern recognition

Given an observation matrix $\mathbf{X}$:

1. Calculate similarity distances d_{ij} between all pairs of snapshots $i, j = 1, \ldots, N$.
2. Threshold the pairwise distances by a suitable kernel of bandwith ϵ and build matrix $\mathbf{D}$ in (3.34).
3. Choose a threshold value T_{shrd}. Construct a truncated distance matrix, $\hat{\mathbf{D}}$, using the following rule:

$$\hat{\mathbf{D}} = \begin{cases} 0 & \text{if } D_{ij} > T_{shrd} \\ D_{ij} & \text{if } D_{ij} < T_{shrd} \end{cases}$$

4. Compute the modified Markov matrix $\mathbf{M}$.
5. Compute the eigenvalues and eigenvectors of the Markov matrix and obtain the diffusion map embedding $[\lambda_1\Phi_1 \quad \lambda_2\Phi_2 \quad \cdots \quad \lambda_k\Phi_k]^T$.
6. Project the diffusion map back into the physical space as $a_j(t) = \mathbf{X}\boldsymbol{\Psi}_j, \quad j = 1, \ldots, k$, and compute time-domain centroids.
7. Cluster the observed oscillations using the k-mean clustering approach or other suitable technique.
8. Obtain related information such as modal damping from the resulting time coefficients.

The reconstructed time story obtained employing the first p modes is obtained as follows:

$$\hat{\mathbf{x}}(t) = \sum_{j=1}^{p} \hat{\mathbf{x}}_j(t) = \underbrace{\lambda_1\mathbf{X}\boldsymbol{\Phi}_1}_{trend} + \underbrace{\lambda_2\mathbf{X}\boldsymbol{\Phi}_2}_{\substack{Oscillatory \\ process\ 1}} + \cdots + \underbrace{\lambda_p\mathbf{X}\boldsymbol{\Phi}_p}_{\substack{Oscillatory \\ process\ p}} \tag{6.11}$$

where the $\hat{\mathbf{x}}_j(t)$ are the components building up the total response, and p is the number of relevant components. As suggested in (6.11) the first component $\hat{\mathbf{x}}(t)$ captures the signal trend; the remaining components are essentially associated with oscillatory behavior.

Two basic approaches to determine the intrinsic dimensionality of the diffusion process have so far been investigated:

1. The presence of a spectral gaps in the eigenvalue spectrum of the matrix $\mathbf{M}$, such that $\lambda_{k+1} \gg \lambda_k$
2. The use of energy measures (refer to Chapter 3)

It follows from (6.11) that

$$\mathbf{x}_{osc}(t) = \mathbf{x}(t) - \mathbf{X}\boldsymbol{\Psi}_1 \tag{6.12}$$

Pattern recognition techniques naturally call for a clustering procedure as a next step.

6.6.2 Data clustering

Machine learning techniques have the potential to be used for classification and regression of measured data. Among other algorithms, *k*-means algorithms can be used to cluster system trajectories.

6.6.2.1 *k*-Nearest neighbors

Let $\mathbf{X} = [\lambda_2\Phi_2 \quad \lambda_3\Phi_3 \quad \cdots \quad \lambda_p\Phi_p]^T$ be the diffusion coordinates to be clustered into a set of K clusters (the number of neighbors), $\mathrm{C} = \{c_1, c_2, \ldots, c_K\}$. The *k*-means algorithm finds a partition such that the squared error between the empirical mean of a cluster and the points in the cluster is minimized. This partition can take many forms, of which the simplest is to estimate the output as the average of the neighbors.

More formally, the squared error between the centroid u_k of a cluster, c_K, and the points x_i in the cluster is defined as the functional [32, 33]

$$J(c_k) = \sum_{x_i \in c_k} \|x_i - \mu_k\|^2 \tag{6.13}$$

where the centroid of a cluster is computed by averaging the coordinates of the objects in a group.

The goal of *k*-means partitioning is to minimize the sum of the squared errors, over all the K clusters

$$J(C) = \sum_{k=1}^{K} \sum_{x_i \in c_k} \|x_i - \mu_k\|^2 \tag{6.14}$$

While simple, the method is found to provide meaningful results when applied to measured data.

6.6.2.2 Computational issues

The main steps of the *k*-mean algorithm can be summarized as follows [32, 33]:

1. Select an initial partition with K clusters.
2. Repeat steps 3 and 4 until cluster membership converges to a given pattern.
3. Generate a new partition by assigning each pattern to its closest cluster center.
4. Compute new cluster centers.

The main shortcoming of this technique is that the number of clusters has to be fixed *a priori*.

6.6.3 Numerical example

To verify the applicability of the developed algorithms, the high-dimensional pattern recognition technique was applied to the simulated data in Figure 6.10. Table 6.4 shows the extracted eigenvalues, while Table 6.5 shows the extracted clusters from the diffusion map in step 4 above.

Figure 6.12 gives the corresponding coefficients $a_o(t)$, $a_1(t)$, and $a_2(t)$.

Table 6.4 Energy contained in the diffusion map coordinates

Temporal coefficient, a_j	Eigenvalue
1	1.0000
2	0.6297
3	0.4165
4	0.2350
5	0.1153
6	0.0722

Table 6.5 k-Means clusters

Cluster	PMU
1	11, 12
2	2
3	10, 7, 4, 5, 6
4	3

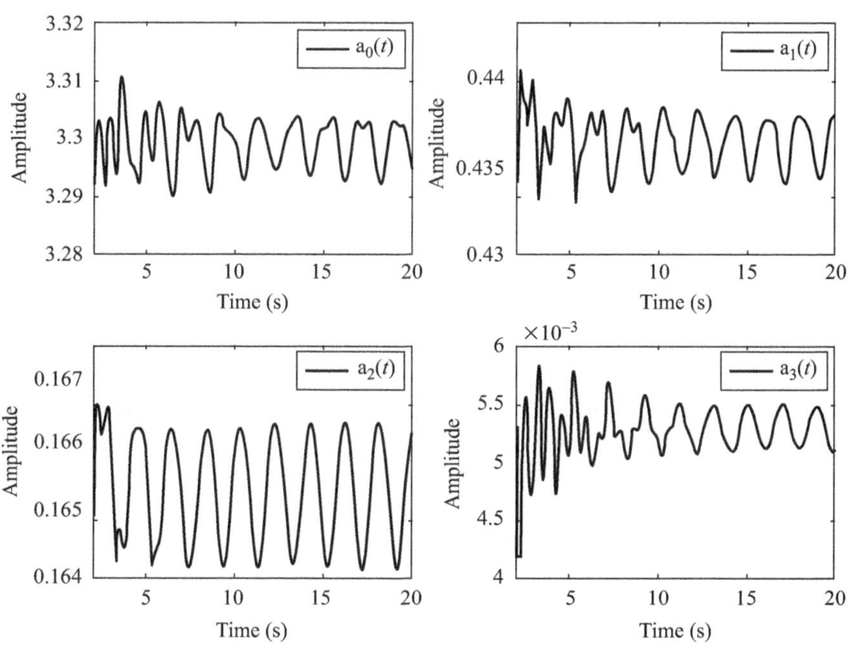

Figure 6.12 Extracting time-varying components using diffusion analysis

Table 6.6 Prony results on diffusion map coefficients

Diffusion coefficient	Frequency (Hz)	Damping ($\xi/2\pi$)
	0.510	−0.00048
a_o	0.917	0.02340
	1.500	0.04260
	0.510	−0.00050
a_1	0.917	0.02340
	1.499	0.04460

The ability of the proposed techniques to preserve the geometric structure of the data is shown in Table 6.6, which displays Prony results of the temporal coefficients. Comparison of modal estimates in Table 6.2 with the modal estimates in Table 6.6 shows that the method accurately preserves system dynamics.

These results are further confirmed in Chapter 8 in the application of these techniques to measured data.

6.6.4 Hybrid schemes

Hybrid schemes combining time-series analysis methods and nonlinear dimensionality reduction techniques provide an alternative to system monitoring.

Several applications are envisaged in the analysis of measured data:

- Identification of geographically homogeneous regions
- Real-time coherency identification
- Modal assessment of power system health

6.7 Voltage and reactive power monitoring

The issue of voltage and reactive power monitoring presents unique challenges. First, dynamic voltage instability processes are characterized by a monotonic voltage drop, which prevents the general application of techniques for oscillatory signals. Second, phase information may be required to characterize reactive power (and voltage) exchange.

Pattern recognition techniques provide a methodology to extract and structure information from large amounts of data and are especially well suited to the analysis of voltage stability and voltage collapse. In this section, the application of these techniques to measured data is examined.

The principles of these approaches are best illustrated with an example.

6.7.1 Measured data

Measured data collected using phasor measurement units (PMUs) in a local 230/400 kV network is used to demonstrate the application of multivariate analysis to the problem of modal identification from voltage and reactive power measurements.

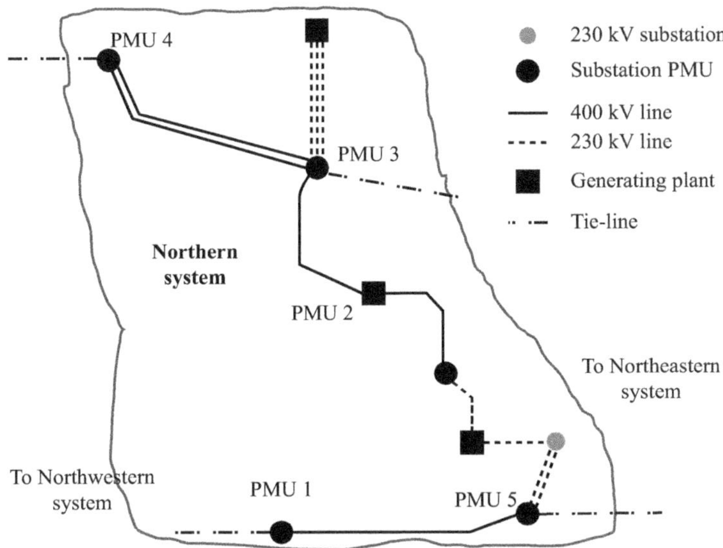

Figure 6.13 Study area showing the location of PMUs and transmission facilities

Figure 6.13 shows the study area depicting the location of five measurement (PMU) sites. Measurements were obtained over a 450 s period at a rate of 20 samples per second.

Bus voltage recordings from selected PMU signals are depicted in Figure 6.14a showing the presence of unstable voltage oscillations. For reference and comparison, Figure 6.14b shows the detrended signals. Note the presence of outliers in the recordings.

Analysis of the power spectra in Figure 6.15 reveals that system behavior is dominated by a mode at about 1.03 Hz. The relatively smaller peaks at 0.41 Hz and 0.63 reveal the presence of major inter-area modes in the system.

Two issues are of concern here: (a) detection of voltage swing deviations from the nominal value and (b) assessment of stability.

6.7.2 Statistical approach to voltage monitoring

Using the above development, time–space analysis is applied to fields of several variables. The voltage-based observation matrix, $\mathbf{X}_v$, is defined as

$$\mathbf{X}_v = \begin{bmatrix} \mathbf{v}_{PMU_1} & \mathbf{v}_{PMU_2} & \cdots & \mathbf{v}_{PMU_5} \end{bmatrix}^T \tag{6.15}$$

where the $\mathbf{v}_{PMU_i}$, $i = 1, \ldots, 5$, represent raw bus voltage measurements at various spatial locations defined as $\mathbf{v}_{PMU_i} = \begin{bmatrix} v_{PMU_i}(t_1) & v_{PMU_i}(t_2) & \cdots & v_{PMU_i}(t_N) \end{bmatrix}^T$.

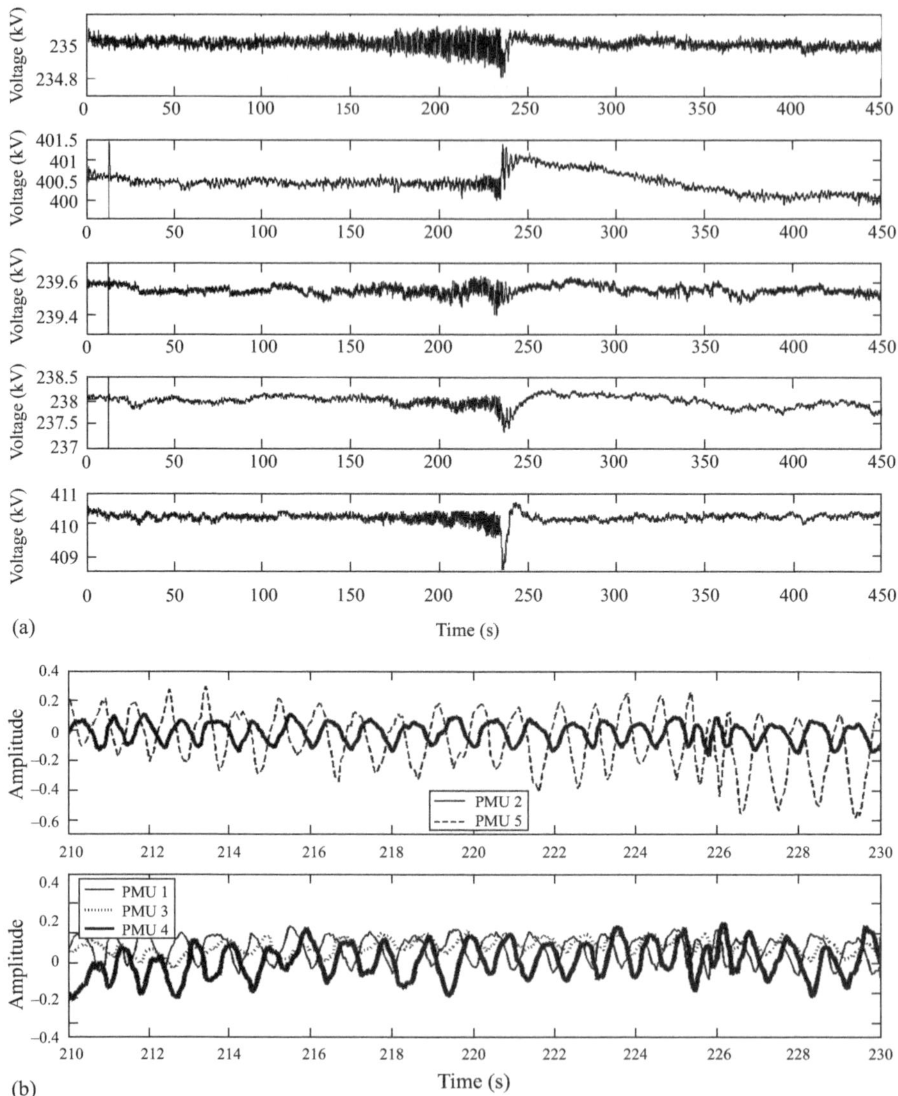

Figure 6.14 Time traces of recorded local voltages. The presence of outliers in the beginning of the data sequence leads to incorrect performance of the estimation methods. (a) Raw bus voltage measurements. Note the presence of outliers in measured data. (b) Detrended signals

Application of multiscale proper orthogonal decomposition (POD) analysis to the data set (6.15) yields a model of the form

$$\mathbf{x}_{v_i}(t_l) \approx \sum_{j=1}^{p} c_{v_{ij}}(t)\,\varphi_{v_j}(\mathbf{x}), \quad i = 1, \ldots, N \tag{6.16}$$

where the $c_{v_{ij}}$ are time-dependent coefficients and the $\varphi_{v_j}(\mathbf{x})$ are spatial coefficients.

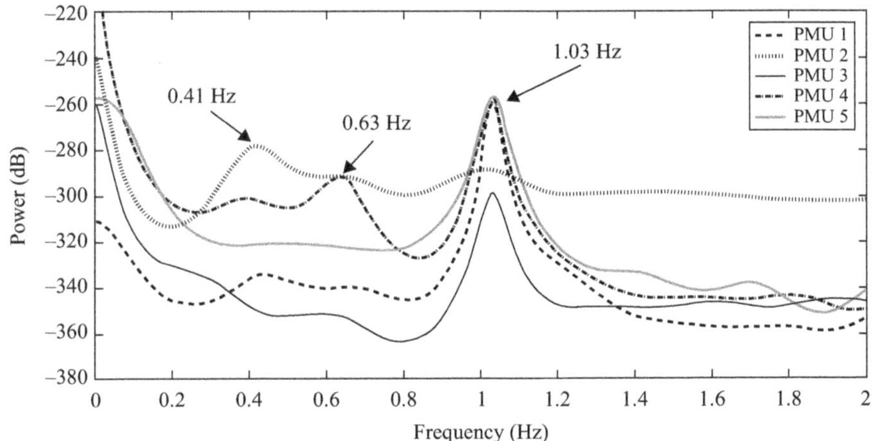

Figure 6.15 Power spectra of voltage measurements in Figure 6.14

Formally, consider a set of measurements described by the spatial and temporal matrix, **X**. Following the general theory in Chapter 3, SVD analysis of the observation matrix gives the linear model

$$\mathbf{X} = \mathbf{\Lambda}_{av} + \mathbf{U}_1\mathbf{\Sigma}_1\mathbf{V}_1^T \tag{6.17}$$

where $\mathbf{\Lambda}_{av}$ represents the time-varying instantaneous means, and the term $\mathbf{U}_1\mathbf{\Sigma}_1\mathbf{V}_1^T$ represents deviations from the mean value.

In the context of the proposed framework, the instantaneous mean deviations can be arranged as a mean deviation feature vector

$$\mathbf{X}_{av} = \begin{bmatrix} x_{mean_1} & x_{mean_2} & \cdots & x_{mean_m} \end{bmatrix}^T \tag{6.18}$$

of the simulated signals, where the x_{mean_k}, $k = 1, \ldots, m$, are the time-varying instantaneous means associated with the kth signal.

For the purposes of this analysis, the nonlinear trends were obtained using the wavelet shrinkage approach. Statistical analysis is then performed on the demeaned matrix $\hat{\mathbf{X}}_v = \mathbf{X}_v - \mathbf{\Lambda}_{av} = \hat{\mathbf{U}}_1\hat{\mathbf{\Sigma}}_1\hat{\mathbf{V}}_1^T$.

Application of the above technique identifies five proper orthogonal modes. Table 6.7 shows the relative energy for each of the POMs, while Table 6.8 shows the extracted modes using Koopman analysis.

Figure 6.16 shows a plot of the time evolution of the first POM. For comparison the time evolution dominant Koopman mode is also plotted.

Also of interest, Figure 6.17 compares the extracted voltage-based mode shape obtained using POD analysis with the Koopman mode operator. Both Koopman analysis and PCA of the dominant mode are found to provide a good estimate to the dominant mode shape, though some differences are noted.

Table 6.7 Singular values of the voltage data

Mode	*Singular value*	Energy (%)
1	0.0234	78.5
2	0.0031	10.4
3	0.0021	7.04
4	0.0011	3.69
5	0.0002	0.67

Table 6.8 Koopman modes

Mode	*Frequency* (Hz)	Damping ($\xi/2\pi$)
1	1.006	−0.029
2	0.627	0.058
3	0.437	0.023

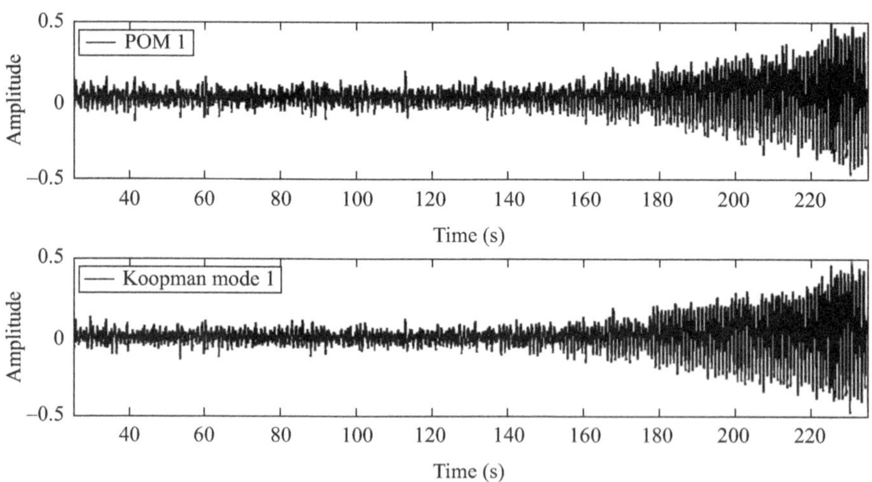

Figure 6.16 Extracted POMs

The analysis achieves the following:

1. Identifies the bus voltage deviation with the largest peak-to-peak-deviation (PMU 5).
2. Identifies bus voltage deviations swinging 180° out of phase for a given mode. As shown in Figure 6.15, voltage deviations at PMUs 1 and 3 are seen to swing in opposition to voltages at PMUs 2, 4, and 5.

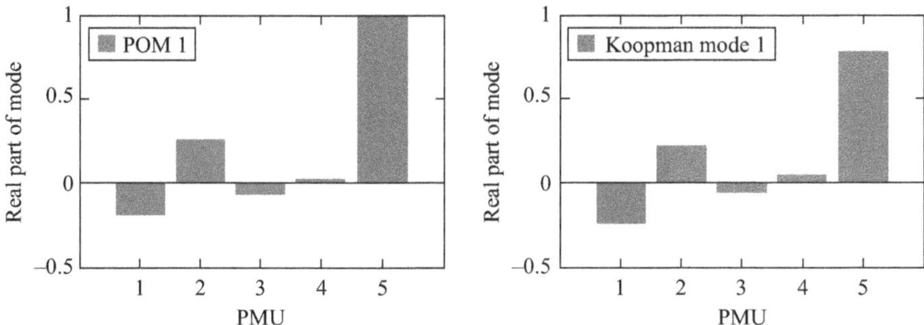

Figure 6.17 Mode shape of 1.0 Hz mode

Table 6.9 Complex mode shape

PMU	Koopman mode 1	POM 1[a]
1	0.4810 ∠–121.67°	0.2948 ∠–133.32°
2	0.5242 ∠58.59°	0.4374 ∠53.32°
3	0.3834 ∠117.58°	0.1670 ∠110.90°
4	0.8966 ∠90.38°	0.4083 ∠85.76°
5	**1.0000 ∠0.0°**	**1.0000 ∠0.0°**

[a]Complex POD formulation.

Results are consistent with the observed time evolution of the signals in Figure 6.14b showing the correctness of the adopted procedure. A drawback of the above approach to analyze measured data is the lack of phase information.

6.7.3 Complex POD/PCA analysis

In order to illustrate the use of complex POD analysis, let the complex observation matrix be defined as $\hat{\mathbf{X}} = \mathbf{X} + jH(\mathbf{X})$ in analogy with complex BSS analysis. Application of POD/PCA analysis results now in complex eigenvalues and eigenvectors.

Table 6.9 compares the complex POD results for mode 1 with the Koopman mode analysis. The analysis of measured data using both Koopman mode analysis and POD analysis in Table 6.9 shows that the bus voltage magnitude at PMU 1 swings in opposition to the bus voltage magnitudes at buses 2–5. Some inconsistencies arise from the use of linear analysis.

As revealed in Table 6.9, POD/PCA analysis tends to underestimate the strength of the modal swings at the PMU locations, especially for PMU 4. Similar results are obtained using the complex BSS representation.

By incorporating phase information, the accuracy of the modal estimation technique is greatly enhanced and the results are more meaningful.

References

1. Georgios B. Giannakis, Vassilis Kekatos, Nikolaos Gatsis, Seung-Jun Hao Zhu, Bruce F. Wollenberg, 'Monitoring and optimization for power grids: A signal processing perspective', *IEEE Signal Processing Magazine*, vol. 30, no. 5, September 2013, pp. 107–128.
2. J. Hauer, D. J. Trudnowski, J. G. DeSteese, 'A perspective on WAMS analysis tools for tracking of oscillatory dynamics', 2007 IEEE Power Engineering Society General Meeting, Tampa, FL.
3. Mladen Kezunovic, Sakis Meliopoulos, Vaithianathan Venkatasubramanian, Vijay Vittal, *Applications of Time-Synchronized Measurements in Power Transmission Networks*, Power Electronics and Power Systems Series, Springer, Cham, Switzerland, 2014.
4. Om P. Dahal, Sukumar M. Brahma, 'Preliminary work to classify the disturbance events recorded by phasor measurement units', 2012 IEEE Power Engineering Society General Meeting.
5. A. Bhyhovsky, J. H. Chow, 'Power system disturbance identification from recorded dynamic data at the northfield substation', *International Journal of Electrical Power Energy Systems*, vol. 25, no. 10, 2003, pp. 787–795.
6. Desiree Phillips, Thomas Overbye, 'Distribution system event detection and classification using local voltage measurements', 2014 Power and Energy Conference at Illinois (PECI).
7. Arturo R. Messina (ed.), *Inter-Area Oscillations in Power Systems*, Springer, New York, NY, 2014.
8. Charles R. Farrar, Keith Worden, *Structural Health Monitoring: A Machine Learning Perspective*, John Wiley & Sons, Ltd, Chichester, West Sussex, UK, 2013.
9. Z. Huang, N. Zhou, F. K. Tuffner, Y. Chen, D. Trudnowski, R. Diao, J. C. Fuller, … J. E. Dagle, *MANGO – Modal Analysis for Grid Operation: A Method for Damping Improvement through Operating Point Adjustment*, PNNL-19890, Pacific Northwest National Laboratory, Richland, WA, October 2010.
10. W. J. Staszewski, C. Boller, G. R. Tomlinson, *Health Monitoring of Aerospace Structures Smart Sensor Technologies and Signal Processing*, John Wiley & Sons, Chichester, West Sussex, UK, 2004.
11. David B. Bertagnolli, Xiachouan Luo, James W. Ingleson, Joe H. Chow, J. Gregory Allcorn, Mark Kuras, Harish I. Mehta, … James P. Hackett, 'Northeastern US oscillation detection and recording project', Fault and Disturbance Analysis Conference, April 2004.
12. A. Bykhovsky, Joe Chow, 'Power system disturbance identification from recorded dynamic data at the northfield substation', *International Journal of Electrical Power & Energy Systems*, vol. 25, no. 10, 2003, pp. 787–795.
13. Richard P. Schulz, Beverly B. Laios, 'Triggering tradeoffs for recording dynamics', *IEEE Computer Applications in Power*, April 1997, pp. 44–49.

14. Daniel J.- Trudnowski, John W. Pierre, Ning Zhou, John F. Hauer, Manu Parashar, 'Performance of three mode-meter block-processing algorithms for automated dynamic stability assessment', *IEEE Transactions on Power Systems*, vol. 23, no. 2, May 2008, pp. 680–690.
15. Arturo R. Messina, Vijay Vittal, 'Extraction of dynamic patterns from wide-area measurements using empirical orthogonal functions', *IEEE Transactions on Power Systems*, vol. 22, no. 2, May 2007, pp. 682–692.
16. Desiree Phillips, Thomas Overbye, 'Distribution system event detection and classification using local voltage measurements', Power and Energy Conference at Illinois (PECI), 2014.
17. Sidharth Thakur, Aranya Chakrabortty, 'Multidimensional wide-area visualization of power system dynamics using synchrophasors', 2013 IEEE Power Engineering Society General Meeting, Vancouver, BC, 2013.
18. Arturo R. Messina, Vijay Vittal, Gerald T. Heydt, Timothy J. Browne, 'Nonstationary approaches to trend identification and denoising of measured power system oscillations', *IEEE Transactions on Power Systems*, vol. 24, no. 4, November 2009, pp. 1798–1807.
19. K. C. Ong, Zenrong Wang, M. Maalej, 'Adaptive magnitude spectrum algorithm for Hilbert–Huang transform based frequency identification', *Engineering Structures*, vol. 30, 2008, pp. 33–41.
20. Fei Bao, Xinlong Wang, Zhiyong Tao, Qingfu Wang, Shuanping Du, 'EMD-based extraction of modulated cavitation noise', *Mechanical Systems and Signal Processing*, vol. 24, 2010, pp. 2124–2136.
21. Nadine Aubry, Régis Guyonnet, Ricardo Lima, 'Spatiotemporal analysis of complex systems: theory and practice', *Journal of Statistical Physics*, vol. 64, nos. 3/4, 1993, pp. 683–739.
22. Nilanjan Senroy, 'Generator coherency using the Hilbert–Huang transform', *IEEE Transactions on Power Systems*, vol. 23, no. 4, November 2008, pp. 1701–1708.
23. Davood Rezaei and Farid Taheri, 'Experimental validation of a novel structural damage detection method based on empirical mode decomposition', *Smart Materials and Structures*, vol. 18, no. 4, 2009, pp. 1–14.
24. J. Schwarz, K. Brauer, G. Dangelmayr, A. Stevens, 'Low-dimensional dynamic and bifurcation in oscillation networks via bi-orthogonal spectral decomposition', *Journal of Physics A: Mathematical and General*, vol. 33, 2000, pp. 3555–3566.
25. Jian Huang, Xiaoguangh Hu, Xing Geng, 'An intelligent fault diagnosis method of high voltage circuit breaker based on improved EMD energy entropy and multi-class support vector machine', *Electric Power Systems Research*, vol. 81, 2011, pp. 400–407.
26. D. J. Trudnowski, J. M. Johnson, J. F. Hauer, 'Making Prony analysis more accurate using multiple signals', *IEEE Transactions on Power Systems*, vol. 14, no. 1, February 1999, pp. 226–231.
27. Power System Dynamic Performance Committee, Task Force on Identification of Electromechanical Modes, Chair: Juan J. Sánchez Gasca,

'Identification of electromechanical modes in power systems', IEEE/PES Special Publication TP462, June 2012.
28. Yoshihiko Susuki, Igor Mezic, 'Nonlinear Koopman modes and coherency identification of coupled swing dynamics', *IEEE Transactions on Power Systems*, vol. 26, no. 4, 2011, pp. 1894–1904.
29. Yoshihiko Susuki, Igor Mezic, 'Nonlinear Koopman modes and power system stability assessment without models', *IEEE Transactions on Power Systems*, vol. 29, no. 2, 2014, pp. 899–907.
30. E. Barocio, Bikash C. Pal, Nina F. Thornhill, A. R. Messina, 'A dynamic mode decomposition framework for global power system oscillation analysis', *IEEE Transactions on Power Systems* (In Press).
31. Vijay Vittal, Trevor Werho, Mladen Kezunovic, Ce Zheng, Vuk Malbasa, Junshan Zhang, Miao He, *Data Mining to Characterize Signatures of Impending System Events or Performance from PMU Measurements*, Final Project Report, PSERC Publication, 13–39, August 2013.
32. Pierre Legendre, Louis Legendre, *Numerical Ecology*, Elsevier Science, Amsterdam, The Netherlands, 1998.
33. Anil K. Jain, 'Data clustering: 50 years beyond *k*-means', 19th International Conference on Pattern Recognition (ICPR), Tampa, FL, December 8, 2008.

Chapter 7
Near real-time analysis and monitoring

7.1 Introduction

Timely and accurate monitoring of system dynamic behavior is essential to improve wide-area situational awareness and system reliability.

In the preceding chapter attention was directed to the formulation of analytic methods to assess power system health. This chapter examines the use of near real-time analysis techniques to detect, locate, and characterize power system disturbances and monitor power system oscillatory dynamics.

Techniques to detrend and denoise measured power system data are outlined and tested on phasor measurement unit (PMU) data. The critical issues for future research in the area of damage identification are also discussed and tools for real-time visualization and monitoring are also reviewed. Emphasis is placed on the development of multiscale, multivariate data analysis techniques.

Examples are used throughout to illustrate various points.

7.2 Toward near real-time monitoring of system behavior

Near real-time monitoring systems have been recently developed and implemented in many power systems. These techniques can be classed into either of the following [1–3]:

1. Block processing techniques
2. Recursive processing techniques

Each class has its own advantages and disadvantages in addressing power system problems. Examples of the first class include techniques such as Prony methods, Hilbert–Huang transform (HHT), autoregressive-moving-average (ARMA), the state-space identification method, and wavelet-based analysis, among other techniques [1, 4–6]. Typical examples of recursive processing techniques, on the other hand, include Kalman filtering, recursive least-squares (RLS) formulations, and hybrid state-space formulations [6–9]. References [4, 10] describe recent applications of these methodologies to analyze power system behavior.

Before discussing these methods, the issues of pre-processing raw data are discussed.

7.3 Data processing and conditioning

Complex oscillatory processes are known to contain noise, trends, and other artifacts that can prevent the analysis and extraction of special features of interest, such as localized events in time [11]. Nontypical behavior can mask out transient oscillations or result in false alarms being sent to the wide-area monitoring systems (WAMS) [12]. Filtering may also be relevant to trigger algorithms [13].

In sections 7.3.1 and 7.3.2, two approaches for denoising are outlined and considered, wavelet analysis and a EMD-based denoising and filtering technique.

7.3.1 Wavelet denoising and filtering

Wavelet shrinkage has recently emerged as a useful signal processing tool for recovering signals from noisy observations. The key idea behind this approach is to take the discrete wavelet transform of the data and shrink or remove the wavelet coefficients to remove the noise. This process is called thresholding.

Following Donoho and Johnstone [14] and Messina *et al.* [11], consider the problem of recovering a function $f(t)$ from noise contaminated observations

$$y_i(t_i) = f(t_i) + \varepsilon(t_i), \quad i = 1, \ldots, N \tag{7.1}$$

with $N = 2^{j+1}$, where y_i is the observed data point, ε is the white noise of unknown variance σ^2, the t_i are equally spaced points, and f is an unknown function to be recovered from the observations.

There are N equations of the form (7.1). Using matrix notation, the discrete wavelet transform of the data can be expressed as

$$\mathbf{w} = \mathbf{W}\mathbf{y} = \mathbf{W}\mathbf{f} + \mathbf{W}\boldsymbol{\varepsilon} \tag{7.2}$$

where $\mathbf{y} = [y_1, \ldots, y_N]^T$ denotes the N-by-1 vector of noisy data, $\mathbf{x} = [x_1, \ldots, x_N]^T$ denotes the vector of unknown signal measurements, $\boldsymbol{\varepsilon} = [\varepsilon_1, \ldots, \varepsilon_N]^T$ denotes the vector of noise, $\mathbf{w} = [w_1, w_2, \ldots, w_N]^T$ denotes the N-by-1 vector of wavelet coefficients of the data, and $\mathbf{W}$ denotes the $N \times N$ orthonormal wavelet transform matrix that contains the filter coefficients defined as

$$\mathbf{W} = \begin{bmatrix} w_{11} & w_{12} & \cdots & w_{1N} \\ w_{21} & w_{22} & \cdots & w_{2N} \\ \vdots & \vdots & \ddots & \vdots \\ w_{N1} & w_{N2} & \cdots & w_{NN} \end{bmatrix} \tag{7.3}$$

This yields the wavelet coefficients w_{jk} with $j = j_o, \ldots, J-1$ and $k = 0, \ldots, 2^j - 1$. In this description, 0 contains the mother and father wavelets; increasing values of J describe finer detail. The problem now becomes that of estimating f from the noisy data, with small mean-square error.

Assume to this end that the wavelet coefficients are modified by some procedure and that an array $\hat{\mathbf{w}}$ is formed. An estimate $\hat{\mathbf{f}}$ of the observed data f at time instants t_i is then obtained as

$$\hat{\mathbf{f}} = \mathbf{W}^T \hat{\mathbf{w}} \tag{7.4}$$

where $\hat{\mathbf{f}} = \left[\hat{f}_1 \ \ \hat{f}_2 \ \ \cdots \ \ \hat{f}_N \right]^T$ is the unknown vector of interest. A standard approach to denoising is to set to zero all coefficients $\hat{w}_{jk}$, $j > J$ below a given threshold u.

Figure 7.1 shows a schematic representation of the filtering process. The whole process can be summarized as follows:

Pseudo-algorithm for wavelet shrinking or denoising

1. Compute the one-dimensional wavelet transform of the data $\mathbf{w} = \mathbf{W}\mathbf{y}$. This is equivalent to decomposing the time series into a linear combination of discrete wavelet orthonormal bases.
2. Estimate the standard deviation of the noise, $\hat{\sigma}$, at level j from $j = 1, \ldots, J$.
3. Apply a thresholding function $\eta_t(y) = \mathrm{sgn}(y)(\|y\| - t)_+$, with the threshold $u = \sqrt{2}\log(N)\sigma/\log(N)$. Set coefficients smaller than the threshold to zero.
4. Reconstruct a denoised version of the original signal, w_k, using the inverse wavelet transform.

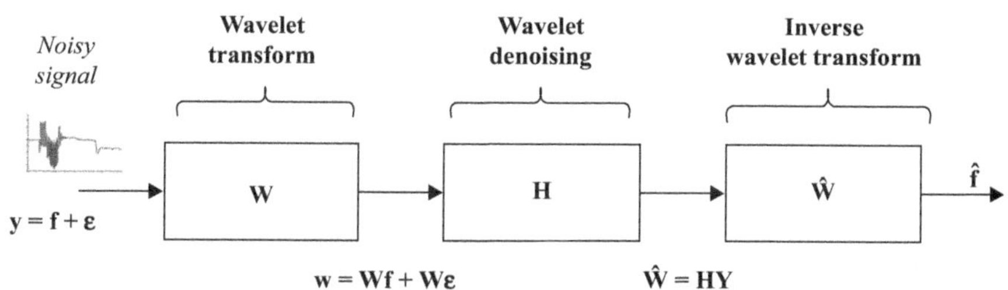

Figure 7.1 Wavelet-based shrinking

The efficiency of the denoising procedure depends on various factors such as the wavelet basis and the threshold selection method.

While this approach can be applied directly to denoise each component of (7.1), the method is not robust since it ignores the correlation structure. A critical step is the selection of the threshold u and the computation of the standard deviation of the noise. In general, automated adjustment of the change detection threshold is required while minimizing the rate of false alarms.

Various strategies, such as median filtering [4], multiparametric analysis [1, 5, 6], machine learning [7–10], and signal quality assessment techniques [11] are used to reduce false alarms. All of these techniques, however, have various limitations. Extensions to the multivariate case are described below.

7.3.1.1 Multivariate wavelet denoising

Wavelet shrinkage has the potential to be applied for smoothing and detrending in real-time applications involving multiple data sets. Consider an $m \times N$ matrix of measurements, $\mathbf{X}$. From (7.1) consider the multivariate model

$$\mathbf{y}(t) = \mathbf{f}(t) + \boldsymbol{\varepsilon}(t) \tag{7.5}$$

where $\mathbf{y}(t)$ and $\mathbf{f}(t)$ are m-dimensional column vectors with elements, $y(t_i)$ and $f(t_i)$, respectively; $\boldsymbol{\varepsilon}(t)$ is an m-dimensional vector of centered white noise with variance Σ_ε.

The multivariate wavelet denoising procedure involves five main steps [15]:

1. Compute the wavelet decomposition up to level J of each row of $\mathbf{X}$. This step produces $J+1$ matrices $D_1, D_2, \ldots, D_J$ containing the details of coefficients from level 1 to J of the p signals.
2. Define a matrix of the noise covariance matrix Σ_ε. Compute the singular value decomposition (SVD) decomposition of $\Sigma_\varepsilon = \mathbf{U}\mathbf{\Lambda}\mathbf{V}^T$.
3. Perform a change of basis $D_j\mathbf{V}$, $1 \leq j \leq J$.
4. Apply the m-dimensional threshold $u_i = \sqrt{2\log(N)}$ to the ith column of $D_j\mathbf{V}$.
5. Using the thresholded information in step 4, reconstruct a denoise matrix $\hat{\mathbf{X}}$ by inverting the wavelet transform.

This results in the approximation

$$\mathbf{y}(t) = \mathbf{f}(t) + \boldsymbol{\varepsilon}(t)$$

that generalizes (7.1) to the multivariate case.

7.3.2 EMD-based filtering

An interesting alternative to nonlinear filtering is based on the empirical mode decomposition technique in Chapter 4.

Based on the theoretical analysis of nonlinear time-series methods in Chapter 6, assume that a measured signal is decomposed in the form

$$x(t) = \underbrace{\sum_{j=1}^{p} c_j(t)}_{\text{Noise+HFC}} + \underbrace{\sum_{k=p+1}^{r} c_k(t)}_{\text{Physically meaningful components}} + \underbrace{\sum_{l=r+1}^{n} c_l(t)}_{\text{Artificial components}} \tag{7.6}$$

where $c_j(t)$ is the jth modal component associated with the frequency ω_j.

The detrended (denoised) signal can now be expressed as

$$\hat{x}(t) = x(t) - \sum_{j=1}^{p} c_j(t) - \sum_{l=r+1}^{n} c_l(t) \tag{7.7}$$

where $\hat{x}(t)$ is the filtered signal and the index r represents a subset of the modal components obtained by discarding nonimportant or uninteresting components in (7.6).

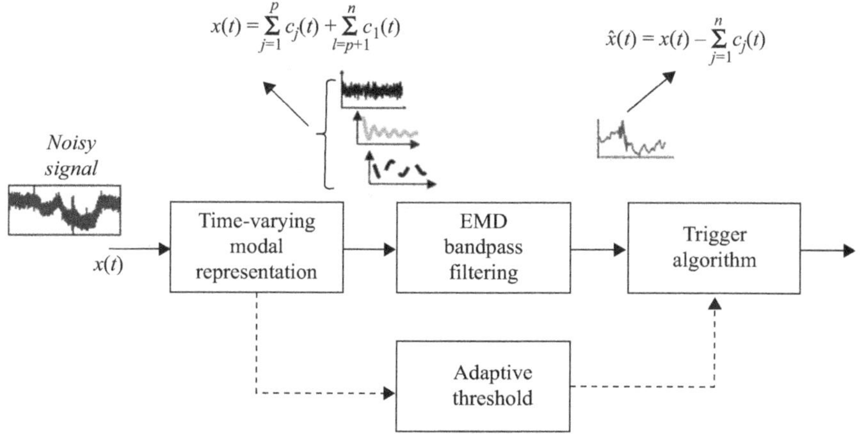

Figure 7.2 Nonlinear EMD filtering

Ideally, the bandpass signal contains oscillatory components associated with a given frequency band of interest.

Figure 7.2 shows schematically the application of this idea. Following the filtering process, a trigger algorithm is used to generate alarms and control actions.

Although the above approach can handle nonlinearities and nonstationarities, their application to a real-time setting is not straightforward.

Extensions to these approaches for near real-time applications are now discussed in the context of existing triggering algorithms.

7.4 Damage detection from changes in system behavior

Damage and disturbance detection is critical for the risk-assessment process. The need for additional or complementary global damage detection methods has led to the development of advanced methods that can assess damage directly from the observed response. These methods should identify damage at an early stage, locate the disturbance location, and provide some estimate of the severity of the damage. The methods should also be well suited to automation.

Several different event detection schemes have emerged in recent works:

1. Pattern-based event classification
2. Modal-based event classification
3. Triggering algorithms

An interesting extension of time-domain filtering and modal attributes is its use as a power swing detector.

7.4.1 Event trigger

In [16–18] techniques to design triggering algorithms for detecting system transients were developed. Reference [18] describes the practical experience with the implementation of these methods.

Detection of transient behavior involves four main activities:

1. Noise removal
2. Trend extraction
3. Detection of signal's *activity*
4. Detection of signal's *persistency*

These issues are now examined in the context of nonlinear time-series filtering and smoothing.

7.4.2 Event detection based on linear filtering

Typical trigger filters consist of three stages: a delayed system, an M-point moving average stage, and a nonlinear (linear) high-pass filter. Figure 7.3 shows a typical representation of these filters [19].

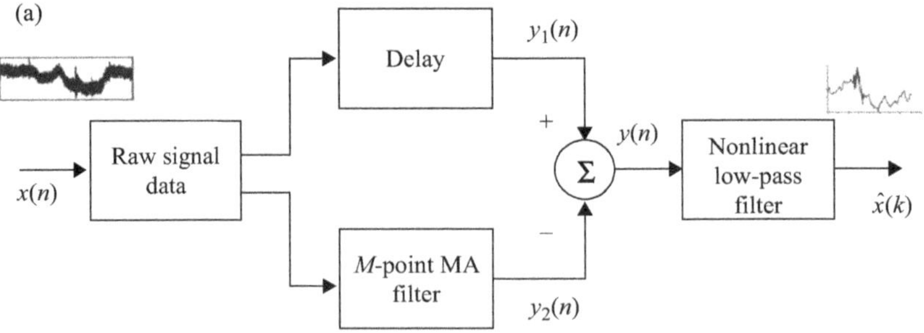

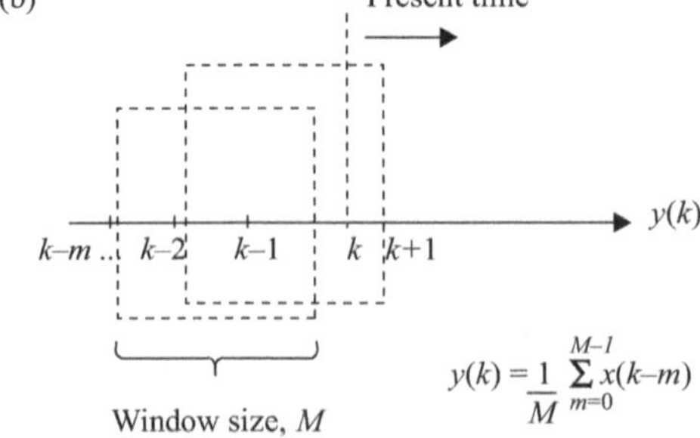

Figure 7.3 Event detection system and its implementation using an M-point average filter. (a) Event detector consisting of a linear high-pass filter and a nonlinear low-pass filter. (b) M-point moving average filter

Referring to Figure 7.3a, the output of the moving average (MA) filter can be expressed as

$$y(n) = \sum_{m=0}^{M-1} b_m x(n-m)$$

where M is the filter length and the b_m are weighting coefficients that define the characteristics of the filter, such that $\sum_m b_m = 1$.

Consider now a delay system with a group delay of $(M-1)/2$ samples, where M is the number of samples in the width of the integration window as shown in Figure 7.3b.

The output of the linear filter can be expressed as

$$y(n) = y_2(n) - y_1(n) = x\left(n - \frac{M-1}{2}\right) - \frac{1}{M}\sum_{m=0}^{M-1} x(n-m) \tag{7.8}$$

References [1–4] discuss the experience in the implementation of these filters. Variants of these strategies are the mean and median filters used to detect system changes in near real-time [20, 21].

7.4.2.1 Online multiscale filtering

A drawback of linear filters is that they represent data at a single scale. As an alternate, wavelet and HHT filtering can be used to sharpen results, especially in the context of complex system oscillations in which data features and noise are not at the same resolution in time and frequency.

Comparison of the models (7.1)–(7.6), and (7.7) and (7.8) leads to the following conclusions:

1. A real-time implementation of a wavelet-based filter can be obtained by adding a wavelet denoising filter to the linear filter in Figure 7.3, as suggested in Figure 7.4.
2. The EMD denoising/filtering process can replace the overall filter in Figure 7.3.

The practical application of these ideas is now tested on measured data.

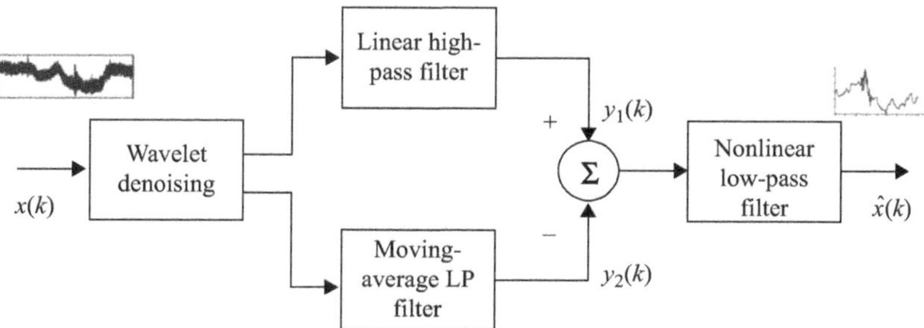

Figure 7.4 Nonlinear inter-area oscillation trigger

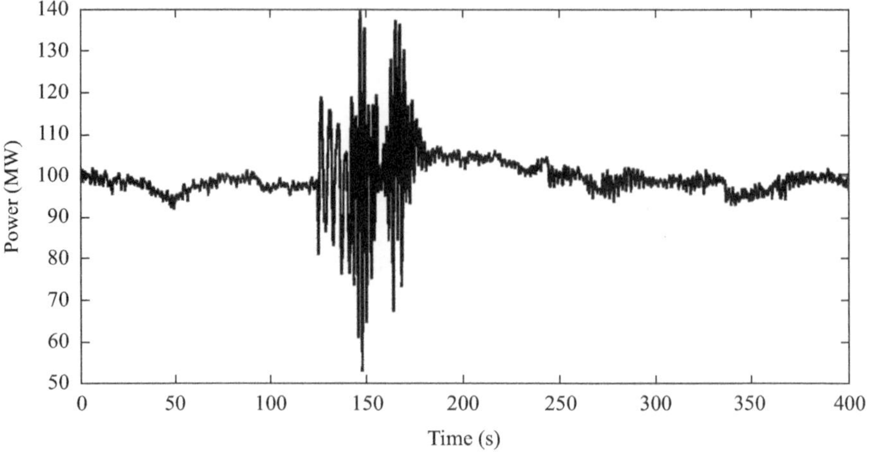

Figure 7.5 Recorded test signal

7.4.3 An illustration

To illustrate the performance of the above approaches, consider the problem of detecting changes in the measured power signal in Figure 7.5. This is the same signal used in the context of nonlinear nonstationary analysis of power system behavior [22].

Figure 7.6 compares the performance of a conventional inter-area oscillation trigger based on Figure 7.3 with the empirical mode decomposition (EMD) filtering technique in Figure 7.1 on measured data. As shown in this plot, both techniques are able to detect the start and end of the oscillations as well as the periods of greater activity. Similar results are obtained using the wavelet-based implementation and are therefore not shown.

A limitation of these approaches, however, is that direct information about the temporal scales associated with the underlying system modes is missing. Also, a multivariate extension to these approaches is deemed necessary since nonlinearity is nonuniformly distributed in the system.

Motivated by these ideas, the following sections explore the use of multiscale statistical approaches to detect abrupt changes and abnormal operation in data series.

The use of variable length sliding window techniques is introduced.

7.5 Time-series approaches to detection of abnormal operation

7.5.1 Near real-time implementations

As discussed above, abnormal operation is detected if the measurements deviate from normal operation. The problem of identifying abnormal operation can thus be formulated as a statistical identification problem, where the scales where significant events are detected are singled out using a nonlinear and/or nonstationary signal processing technique.

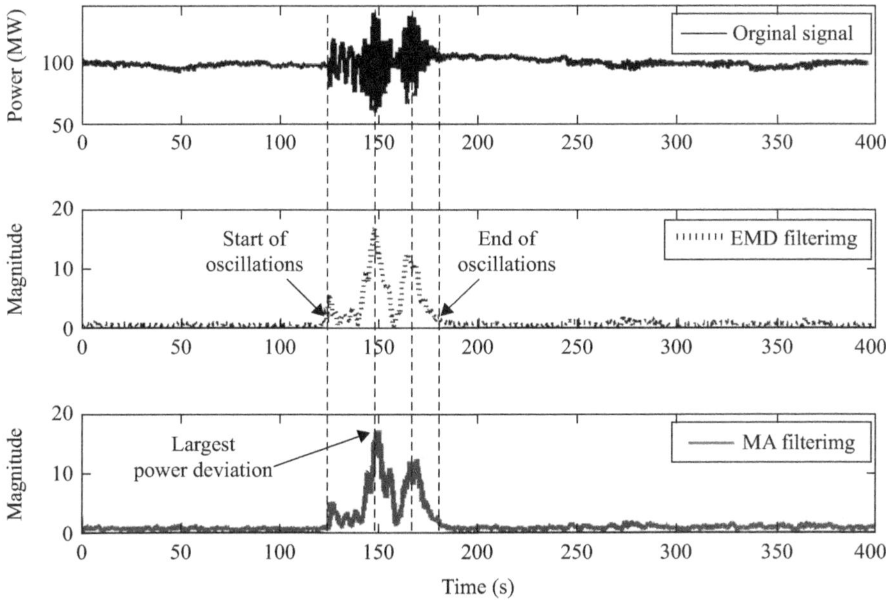

Figure 7.6 Comparison of inter-area oscillation trigger algorithms: (a) raw signal; (b) EMD filtering; (c) MA filtering

Figure 7.7 shows, schematically, the proposed procedure. For clarity of illustration a single time series is shown, but the approach can be applied to several signals recorded simultaneously. In the multivariate case, the observation matrix can be computed sequentially for two adjacent time intervals.

To capture multiscale behavior, the following procedure is adopted:

Pseudo algorithm for multiscale detection of abnormal operation

1. Given a set of simultaneously recorded signals $x_k(t)$, $k = 1, \ldots, m$.
2. Decompose the signal $x(k)$ into frequency components $c_1(t_{Wk}), \ldots, c_r(t_{Wk})$, where t_{Wk} denotes the kth time window using a time–frequency analysis approach.
3. Compute the principal component analysis (PCA) decomposition using the approach in Chapter 6. Alternatively, perform linear (nonlinear) PCA on the raw measurements.
4. For each scale (mode) of interest, assess the change in modal properties.

Abnormal operation at different temporal scales is detected if the measurements deviate from the region of normal space in the retailed principal component scores.

At each time window, the measured data is decomposed in the form

$$\hat{x}(t) = x(t) - \sum_{j=1}^{p} c_j(t) - \sum_{l=r+1}^{n} c_l(t)$$

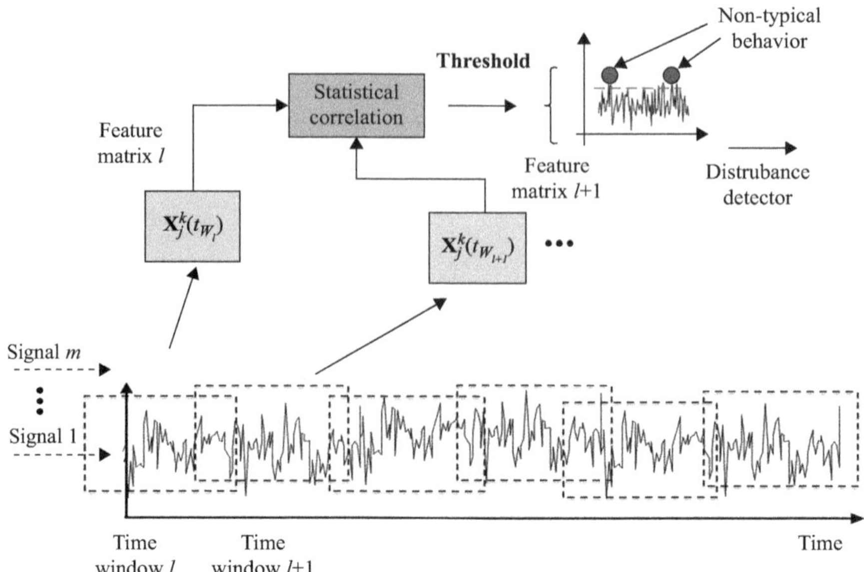

Figure 7.7 Window-based implementation of a disturbance detection scheme. The model can be applied to generate temporal feature matrices $\mathbf{X}_f$ *in a multiway PCA monitoring technique*

As discussed in Chapter 6, the data can be analyzed at sensor or system level. When monitoring of a given modal component is of interest, the multiblock approach can be used. For every segment or time window, the observation matrix can be constructed as follows:

$$\mathbf{X}_j^k(t_{w_k}) = \begin{bmatrix} c_{1j}^k(t_1) & c_{1j}^k(t_2) & \cdots & c_{1j}^k(t_N) \\ c_{2j}^k(t_1) & c_{2j}^k(t_2) & \cdots & c_{2j}^k(t_N) \\ \vdots & \vdots & \ddots & \vdots \\ c_{m_kj}^k(t_1) & c_{m_kj}^k(t_2) & \cdots & c_{m_kj}^k(t_N) \end{bmatrix}, \quad j = 1, \ldots, p_k$$

This yields a feature matrix of the form

$$\mathbf{X}_j(t_{w_k}) = \begin{bmatrix} \mathbf{X}_j^1(t_{w_k}) \\ \hdashline \mathbf{X}_j^2(t_{w_k}) \\ \hdashline \vdots \\ \hdashline \mathbf{X}_j^M(t_{w_k}) \end{bmatrix} \tag{7.9}$$

that results in a three-way decomposition of the data. Figures 7.7 and 7.8 illustrate these concepts.

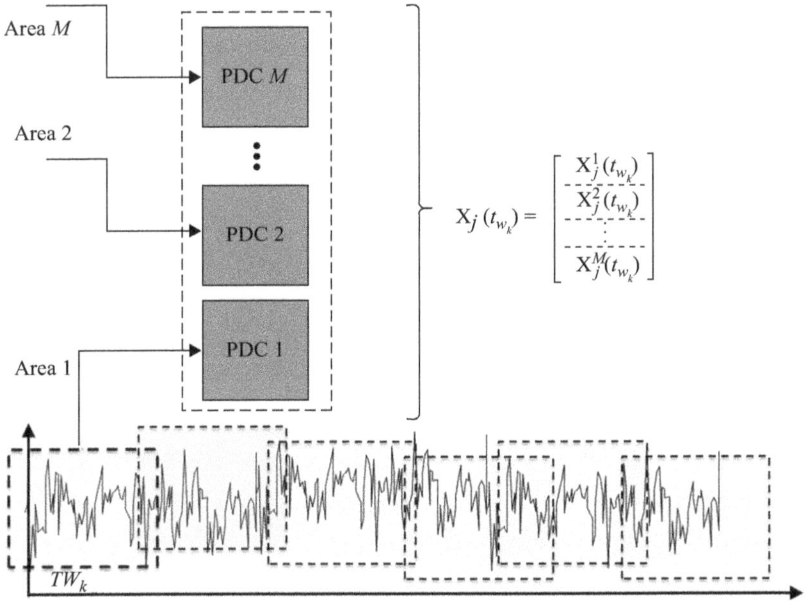

Figure 7.8 Construction of a three-way decomposition of data for an M-area system

The outcome of this approach is disturbance detection at a given scale, which makes it particularly useful for early warning of modal behavior or the use of control actions in the context of recent applications.

7.5.2 Near real-time implementation of the Hilbert transform

The above implementation requires the solution of two distinct problems [23–25]:

1. The computation of a local decomposition of the signal
2. The computation of the instantaneous modal parameters

These issues are now addressed in the context of the HHT. As discussed previously, a local decomposition can be obtained using EMD. Once a local decomposition is obtained, the modal parameters can be derived using various approaches.

There are several approaches to computing the Hilbert transform. Table 7.1 summarizes some methods of interest.

The Hilbert transform of a function $f(x_k)$ is defined as

$$H[f(x_k)] = \frac{1}{\pi}\int_{-\infty}^{\infty}\frac{f(t)}{x - x_k}dx = -\frac{1}{\pi}\int_{-\infty}^{\infty}\frac{f(t)}{x - x_k}dx \tag{7.10}$$

This representation leads to some insights that can be exploited for near real-time analysis. For discrete data, the Hilbert transform can be approximated by a finite series as

$$\int_a^b f(x)dx \approx \sum_{n=1}^{N} w_n f(x_n)$$

where N is the number of sample points and the term w_n represents weighting coefficients at the sample point.

Using trapezoidal integration, the above integral can be approximated as

$$\int_{-\infty}^{\infty} f(x)dx \approx h \sum_{k=-\infty}^{k=\infty} w_n f(kh)$$

where h is the step size of the equally space intervals.

This approximation can be implemented as

$$\int_{-\infty}^{\infty} f(x)dx \approx h \sum_{k=-\infty}^{k=\infty} f(kh + x_n)dx, \qquad -\infty \leq x_n \leq \infty$$

After some manipulations, it can be shown that

$$H[f(x_n)] = \frac{1}{\pi} \int_{-\infty}^{\infty} f(x)dx \approx \frac{2h}{\pi} h \sum_{k=-\infty}^{k=\infty} \frac{f(x_n + kh)}{(x_n + kh) - x_n}$$

or

$$H[f(x_n)] \approx \frac{2}{\pi} \sum_{\substack{k=-\infty \\ k \text{ even}}}^{k=\infty} \frac{f(x_n + kh)}{k}$$

Note now that it can be shown that

$$H[f(x_n)] \approx \frac{2}{\pi} \sum_{k \geq 0}^{\infty} \frac{1}{2k+1} (x_{n+2k+1} - x_{n-2k+1})$$

Variants to this approach have been discussed in the power system literature and are briefly reviewed here. In [26], the Hilbert transform was obtained directly by operating the real component with a convolution filter

$$\hat{x} = x_H(t) = \sum_{k=-M}^{k=M} x(t-k)h(k) \tag{7.11}$$

where $h(.)$ is the convolution filter with unit amplitude response and 90° phase shift. A simple filter that provides an adequate amplitude response and phase response is given by [32] as

$$h(k) = \begin{cases} \dfrac{2}{\pi k}\sin^2(\pi k/2) & \text{if } k \neq 0 \\ 0 & \text{if } k = 0 \end{cases}$$

where $-M < 1 < M$.

Clearly, as $M \rightarrow \infty$, the filter $h(k)$ yields an exact Hilbert transform. For finite M, the filter introduces ripple effects. To limit these effects, a local Hilbert transform has been developed based on filter banks.

As suggested in [25], the filter banks can be developed such that the flatness of the frequency response is maximal for the length of the filter. A maxflat filter can be defined by [25, 26]

$$h(z) = \left(\frac{1 + z^{-1}}{2}\right) Q_{2p-2}(z)$$

where p is the number that determines the zeros at $\omega = \pi$, and Q is chosen such that $h(z)$ is half-band. The filter $h(z)$ is shifted in frequency by $\pi/2$.

Other potential approaches include Chebyshev filters. Table 7.1 summarizes recent approaches to the computation of the Hilbert transform.

7.5.2.1 Instantaneous frequency approximations

In previous sections the notion of instantaneous frequency was introduced based on the Hilbert transform. Following Barnes [27], let $x(t)$ be a measured signal and $x_H(t)$ be its Hilbert transform. The analytic signal $z(t)$ is defined as

$$z(t) = x(t) + jx_h(t) = A(t)e^{j\theta(t)} \tag{7.12}$$

Table 7.1 Computation of the Hilbert transform

Method	Analytic formulation	Reference
Fourier	$h(k) = \begin{cases} -j, & \text{for } j > 0 \\ 0, & \text{for } j > 0 \\ j, & \text{for } j > 0 \end{cases}$	
Convolution	$h(k) = \begin{cases} \frac{2}{\pi k}\sin^2(\pi k/2) & \text{if } k \neq 0 \\ 0 & \text{if } k = 0 \end{cases}$	
Real time	$HT[x(t_k)] = \frac{2}{\pi}\sum_{n\geq 0}^{\infty}\frac{1}{2n+1}(x_{t+2n+1} - x_{t-2n-1}).$	
Filters	Southeastern system	

Having computed the analytic signal, the instantaneous frequency, $f(t)$ is defined as [28]

$$f(t) = \frac{1}{2\pi}\frac{d}{dt}\theta(t) \tag{7.13}$$

In practice, the instantaneous frequency is calculated directly from the analytic signal. Making use of (4.10) it follows that

$$\omega(t) = \mathrm{Im}\frac{\dot{z}(t)}{z(t)} = \mathrm{Im}\frac{\dot{x}(t) + j\dot{x}_H(t)}{x(t) + jx_H(t)} = \frac{x(t)\dot{x}_H(t) - \dot{x}(t)x_H(t)}{x^2(t) + x_H^2(t)} \tag{7.14}$$

where use has been made of (7.12).

Unwrapping the instantaneous phase resolves possible phase shift of $\pm\pi/2$ due to the ambiguity of the *arctan* function. However, this method may deteriorate with increasing noise. Moreover, the computation of instantaneous frequency evaluates two time derivatives and is numerically sensitive to the effects of low-amplitude areas, that is, $x(t) = x_H(t) \approx 0$.

This results in spurious spikes, the ringing effect caused by Gibb's phenomenon, and is quite unstable when the input signal contains more than one frequency component. As a consequence, practical instantaneous frequency estimators need to incorporate some sort of filtering to reduce these effects. Moreover, computation of the instantaneous frequency requires two differentiations.

To motivate the more general ideas that follow, the average instantaneous frequency $f_a(t)$ is defined as the temporal average of instantaneous frequency in a time interval from t to $t + T$, that is

$$f_a(t) = \frac{1}{T}\int_t^{t+T} f(\tau)d\tau \tag{7.15}$$

where τ is a dummy variable of integration.

Numerical alternatives to this formulation that avoid the use of second derivatives are described by Barnes [27] and Lezama [29].

Substitution of (7.13) in (7.15) for $f(t)$ yields

$$f_a(t) = \frac{1}{T}\int_t^{t+T} f(\tau)d\tau = \frac{1}{2\pi T}\int_t^{t+T}\frac{d}{d\tau}\theta(\tau)d\tau = \frac{1}{2\pi T}\frac{\theta(t+T) - \theta(t)}{\Delta t}$$

An interesting alternative expression of the instantaneous frequency is now obtained from the analysis of phase differences directly in the definition of the analytic signal. From (7.12) one has that

$$\theta(t) = \mathrm{Im}[\ln z(t)]$$

Therefore

$$f_a(t) = \frac{1}{2\pi T}[\mathrm{Im}\{\ln z(t + \Delta t)\} - \mathrm{Im}\{\ln z(t)\}]$$

or

$$f(t) = \frac{1}{2\pi T}\left[\arctan\frac{x_H(t+T)}{x(t+T)} - \arctan\frac{x_H(t)}{x(t)}\right]$$
$$= \frac{1}{2\pi T}\frac{x(t)x_H(t+T) - x(t+T)x_H(t+T)}{x(t)x(t+T) + x_H(t)x_H(t+T)} \tag{7.16}$$

where use has been made of the trigonometric identity $\arctan(\alpha) - \arctan(\beta) = \arctan((\alpha - \beta)/(1 + \alpha\beta))$.

This approximation is faster to compute than the definition of instantaneous frequency in (7.14) because it avoids the two differentiations that the computation of instantaneous frequency requires and can be implemented using three data points.

Several other interpretations are possible. Table 7.2 summarizes some alternatives to the computation of instantaneous frequency. Refer to [29] for numerical experience in the use of these models.

So far, however, the experience with the applications of these approximations to power system data has been limited.

7.5.3 Local mean speed

A second approach to detect changes in system behavior is based on the notion of local mean speed.

Figure 7.9 illustrates the decomposition of measured data into its time varying mean and the local mean speed. Several approaches to compute the local mean exist, and its benefits have not been fully explored. Transient detection and segmentation are interrelated.

As shown in the diagram, the trend itself may be a useful indicator of changes in system behavior. The analysis suggests that the slope and the associated local trend can be used to detect change points and anomalies in the observed signal. More precisely, the time instants of change of energy correspond to segmentation boundaries that define segmentation boundaries for nonstationary analysis.

The challenge is to translate this information into criteria to detect abnormal events.

Table 7.2 Computation of the Hilbert transform

Method	**Analytic formulation**
Time difference	$\omega(t) = \left(\frac{1}{2\pi}\right)\frac{d}{dt}\left[\arctan\left(\frac{y(t)}{c(t)}\right)\right] = \left(\frac{1}{2\pi}\right)\frac{c(t)\ddot{c}_H(t) - y(t)\dot{c}(t)}{c^2(t) + c_H^2(t)}$
Phase difference	$\omega(t) = \frac{\Delta\varphi(t)}{\Delta(t)} = \frac{1}{2\pi}\left[\frac{\varphi(t+T) - \varphi(t)}{t+T-t}\right] = \frac{1}{2\pi}\left[\frac{\varphi(t+T) - \varphi(t)}{T}\right]$
Real time	$\omega(t) = \frac{1}{T}\arctan\left[\frac{x(t)x_H(t+T) - u(t+T)x_H(t)}{x(t)u(t+T) + x_H(t)x_H(t+T)}\right]$
Filters	$\omega(t) = \frac{1}{T}\arctan\left[\frac{x(t-T)x_H(t+T) - x(t+T)x_H(t-T)}{x(t-T)u(t+T) + x_H(t+T)x_H(t-T)}\right]$

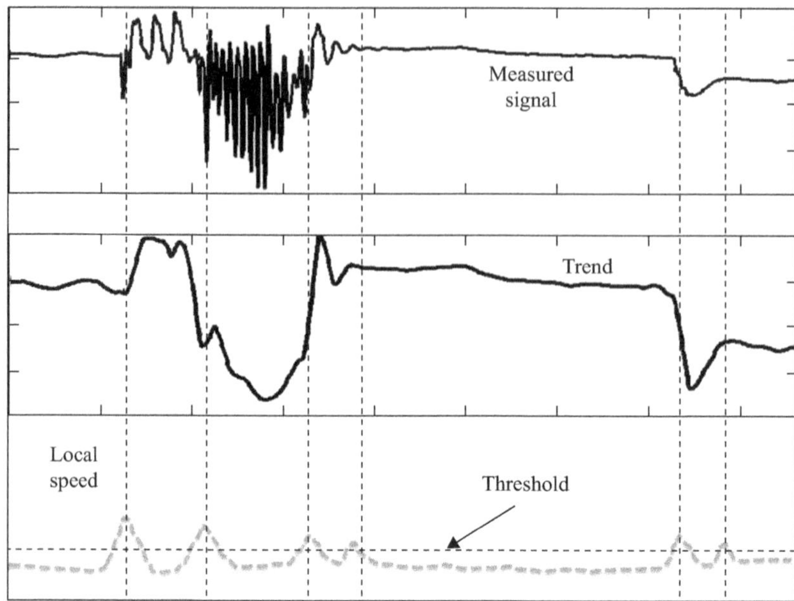

Figure 7.9 Local mean-based implementation of a disturbance detection scheme: (a) measured signal; (b) extracted local mean; (c) segmentation threshold

Next, techniques to determine threshold measures from measured signal are outlined and discussed. The basis for the following discussion will be the EMD procedure detailed in Appendix C, but the adopted approaches extend naturally to other techniques.

With reference to Figure 7.10, define the vector of time instances $\mathbf{t}_l = [t_l^{(1)} \quad t_l^{(2)} \quad \cdots \quad t_l^{(N)}]$ and $\mathbf{t}_u = [t_u^{(1)} \quad t_u^{(2)} \quad \cdots \quad t_u^{(N)}]$ associated with the time instances at which local minima or maxima occur, respectively. Associated with these vectors are the lower and upper values $\mathbf{I}_l = [I_l^{(1)} \quad I_l^{(2)} \quad \cdots \quad I_l^{(N)}]$ and $\mathbf{I}_u = [I_u^{(1)} \quad I_u^{(2)} \quad \cdots \quad I_u^{(N)}]$.

At time instant $t = t_l^{(k)}$, an estimate of the local mean can be obtained from

$$m^{(k)}(t) = \frac{h_u^{(k+1)} - h_l^{(k)}}{t_l^{(k+1)} - t_l^{(k)}} \tag{7.17}$$

and

$$m^{(k+1)}(t) = \frac{h_u^{(k+1)} - h_l^{(k)}}{t_l^{(k+1)} - t_l^{(k)}} \tag{7.18}$$

Combining (7.17) and (7.18) yields

$$\frac{m^{(k+1)}(t)}{m^{(k)}(t)} = \frac{h_u^{(k+1)} - h_l^{(k)}}{h_u^{(k+1)} - h_l^{(k+2)}}$$

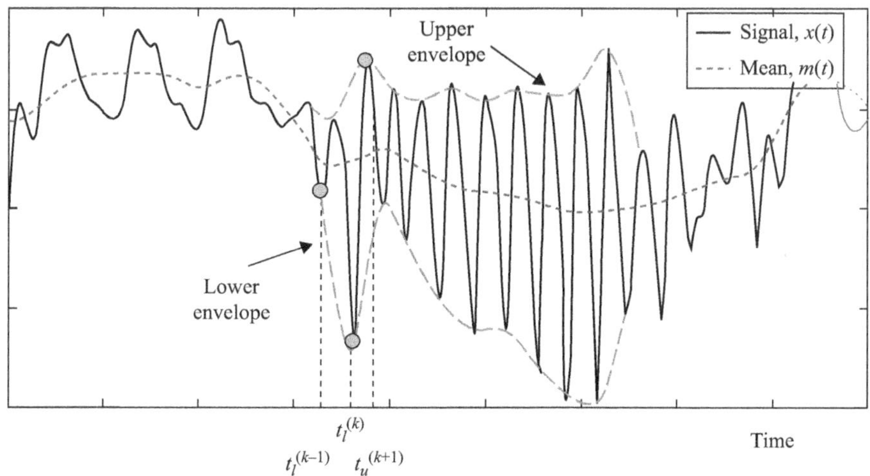

Figure 7.10 Measured signal showing the definition of a local trend

and

$$m_r(k) = \left|\frac{m^{(k+1)}(t)}{m^{(k)}(t)}\right| = \left|\frac{h_u^{(k+1)} - h_l^{(k)}}{h_u^{(k+1)} - h_l^{(k+2)}}\right| \tag{7.19}$$

Finally, the normalized mean ratio can be defined as

$$\hat{m}_r(k) = \frac{MR(k)}{MR_{\max}(k)} \tag{7.20}$$

Using this criterion, a dynamic event is detected if

$$\gamma_{\min}\hat{m}_r \leq \hat{m}_r(k) \leq \hat{m}_r\gamma_{\min} \tag{7.21}$$

where $\gamma_{\min} = 0$ and $\gamma_{\max}$ can be chosen adaptively or on practical or statistical grounds.

In the more general case, the mean can be calculated for a time window, t_k, as

$$m_{t_k} = \frac{1}{\Delta t}\int_{t-\Delta t/2}^{t\Delta t/2} x(t)dt$$

where T is a suitable analysis period or computed rigorously through the procedure outlined in Chapter 4.

These approaches can be used to generate alarms based on criteria such as thresholds, rates of change, and persistence.

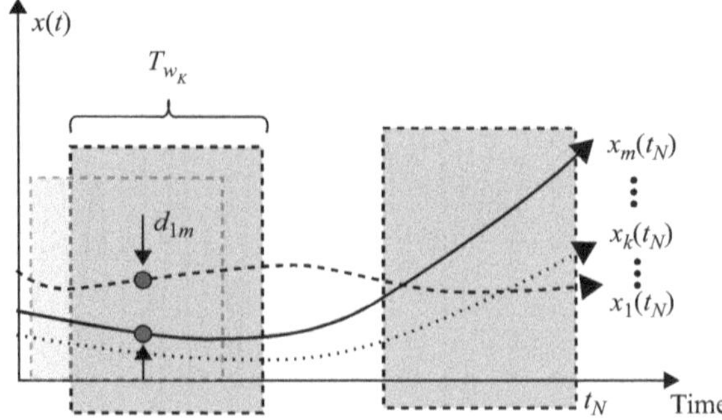

Figure 7.11 An illustration of a sliding window-based implementation of a dimensionality reduction technique for use in pattern recognition of system disturbances

7.6 Pattern recognition-based disturbance detection

Event detection and classification is a natural step in a data fusion scheme and can be posed as a pattern recognition problem.

Consider to this end a set of simultaneous measurements $\mathbf{x}(t_j) = [x_1(t), x_2(t), \ldots, x_m(t)]^T$, $1 \leq t \leq N$. Assume further that the recorded signals are segmented into observation windows of finite length t_W, as shown in Figure 7.11. A sample-by-sample window approach is adopted here but other formulations are possible.

The goal is to identify patterns from the observed measurements. This involves two main steps: (i) extracting from a large set of dynamic trajectories those associated with relevant system behavior and (ii) identifying dynamically relevant patterns.

Assume that system motion is described by a distance matrix **K**. Using the diffusion framework in section 3.5.2, the following approach can be used to identify the onset of system disturbances. Several applications are envisaged. These include but are not limited to: (i) Transient (mid-term) instability detection, (ii) coherency identification, and (iii) modal instability analysis.

It is of note that in the extreme case a sample-by-sample application of the method is possible. Application of the technique is straightforward.

High-dimensional pattern recognition procedure

Given a set of simultaneously recorded signals $x_k(t)$, $k = 1, \ldots, m$

1. Calculate time-dependent similarity distances $d_{ij}(t)$ between all pairs of snapshots i, $j = 1, \ldots, N$.
2. Threshold the pairwise distances by a suitable kernel of bandwidth $\varepsilon, e^{-\|x_i(t)-x_j(t)\|^2/\varepsilon}$, and build an $m \times m$ distance matrix **A**. Construct the corresponding Markov matrix **M**.
3. Calculate diffusion distances.
4. Compute the eigenvalues and eigenvectors of the Markov matrix and define a diffusion map $\mathbf{\Psi} = [\lambda_1\mathbf{\Phi}_1 \quad \lambda_2\mathbf{\Phi}_2 \quad \cdots \quad \lambda_k\mathbf{\Phi}_k]^T$.
5. Project the diffusion map back into the physical space as $\mathbf{a}_j(t) = \mathbf{X}\mathbf{\Psi}_j,\ j = 1, \ldots, k$, and compute time-domain centroids.
6. Cluster the observed oscillations using the k-mean clustering approach or other suitable technique.
7. Obtain related information such as modal damping from the resulting time coefficients.

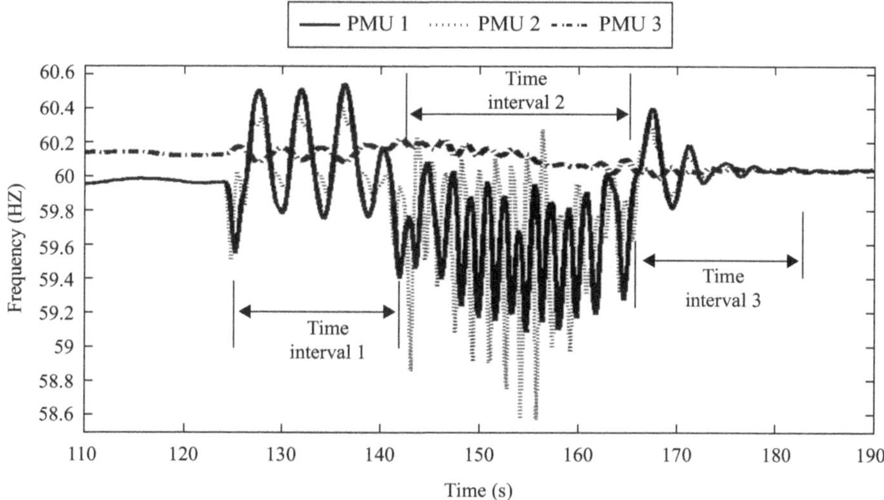

Figure 7.12 Measured signals exhibiting changes in temporal modal behavior [21]

To illustrate this idea, a set of three frequency measurements exhibiting phase dynamic changes is analyzed (refer to Figure 7.12). Three time intervals can be recognized associated with changes in temporal behavior.

For simplicity of discussion, diffusion maps were calculated for nonoverlapping windows of 10 s length. The analysis focuses on the extraction of mode shapes.

Figure 7.13 shows the extracted instantaneous mode shape for selected time intervals as a function of time for three time intervals. The cluster displays several important features:

- *Pattern recognition:* Figure 7.13 captures transient changes in mode shapes.
- *Feature extraction:* Damage sensitive measurement properties are derived. The analysis shows that for time intervals 1 and 3, the frequency measurement associated with PMU 3 has the largest distance.
- *Damage detection:* Changes in mode shape can be used to infer changes in system dynamic behavior.

7.7 Sliding window-based methods

One way to have a block-processing implementation of the above procedures is to use sliding window techniques. In these approaches, a time series is adaptively segmented into observation windows of adjustable length T_W and a sliding window-based method is applied to each window.

7.7.1 Local HHT analysis

Conventional HHT analysis may result in over-decomposition of the signal and other undesirable effects [24]. One way to circumvent these limitations is to define

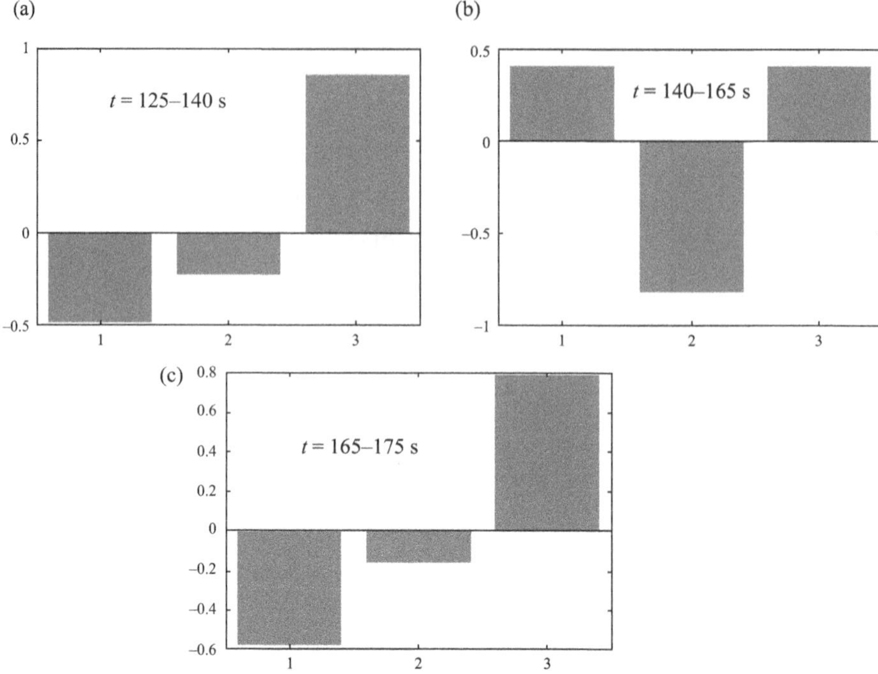

Figure 7.13 Instantaneous mode shape extracted using an online implementation of a diffusion-based pattern recognition technique

a local empirical model decomposition for a sliding window. This can be accomplished block wise without knowledge of the entire signal. More precisely, a sliding window is applied to the signal and the conventional EMD technique is applied to the selected widow.

A generic implementation of the local HHT is shown in Figure 7.14. In the suggested implementation, each set of IMFs is computed by dividing the time series into lengths of time and applying the conventional EMD to the selected window.

Three considerations are introduced in this formulation:

1. The same number of sifting steps is applied to all windows in order to avoid discontinuities.
2. As a first step, it is assumed that there are no overlaps between windows.
3. The number of sifting operations is fixed *a priori*.

It should be emphasized that a local implementation of the Hilbert transform is needed to compute a local estimate of the analytic signal in (7.12). The reader is referred to [2] for a discussion on his subject.

Two approaches are possible. In the first approach, the HHT is applied to a single time series. In the second approach, the signal is adaptively segmented into a number of time windows using a suitable entropy criterion.

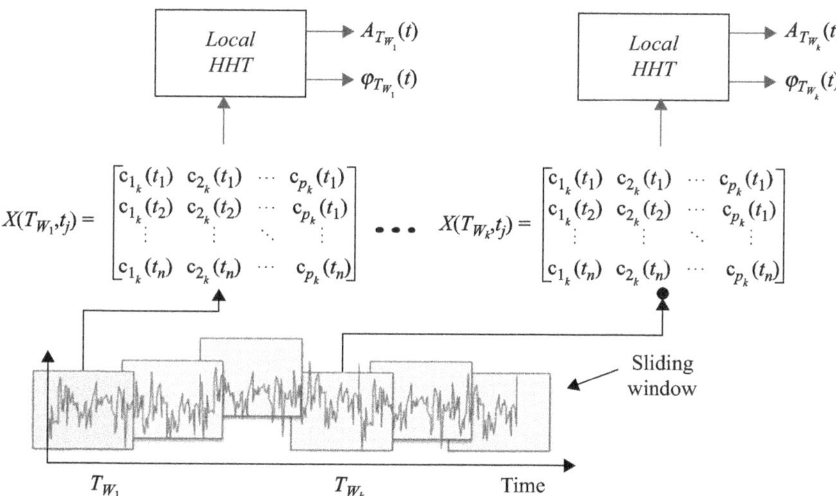

Figure 7.14 Block-wise moving window approach to HHT analysis

The response matrix, **X**, for a given time window t_{Wk} now takes the form

$$\mathbf{X}(m, t_{W_j}) = \begin{bmatrix} c_{1_m}(t_{W_1}) & c_{2_m}(t_{W_1}) & \cdots & c_{n_k}(t_{W_1}) \\ c_{1_m}(t_{W_2}) & c_{2_m}(t_{W_2}) & \cdots & c_{n_k}(t_{W_2}) \\ \vdots & \vdots & \ddots & \vdots \\ c_{1_m}(t_{W_p}) & c_{2_m}(t_{W_p}) & \cdots & c_{n_k}(t_{W_p}) \end{bmatrix} \tag{7.22}$$

In previous research, a sliding window-based approach has been combined with the EMD method to resolve localized information that extends previous research [29]. Several other implementations involving variations to this approach are now introduced and tested on measured data.

By defining a local EMD for a sliding window around a selected time interval, t_W, localized features can be identified and extracted.

Several implementations are possible, including:

1. Overlapping windows
2. Nonoverlapping windows

A number of practical problems have yet to be addressed in implementing the above strategy and are the subject of intense current research. In order to address the above issue and to accommodate data from each of the measurement places, a multivariate statistical control approach has been combined with the HHT technique for monitoring system behavior.

The practical application of this approach is presented in Chapter 8.

7.7.2 Numerical example

An illustration of this idea is provided in Figure 7.15. The top panel shows a plot of the time evolution of recorded active power at a critical interface. Measurements were recorded over 250 s collected at a rate of 20 samples per second for a total of 2 000 samples.

For reference and comparison, conventional (offline) HHT analysis was first applied to extract the dominant IMFs. The results were compared with the local HHT analysis described above. The middle and lower panels in Figure 7.15 compare the reconstructed signal obtained using the local and block-processing technique. A 30 s window size is adopted for illustration.

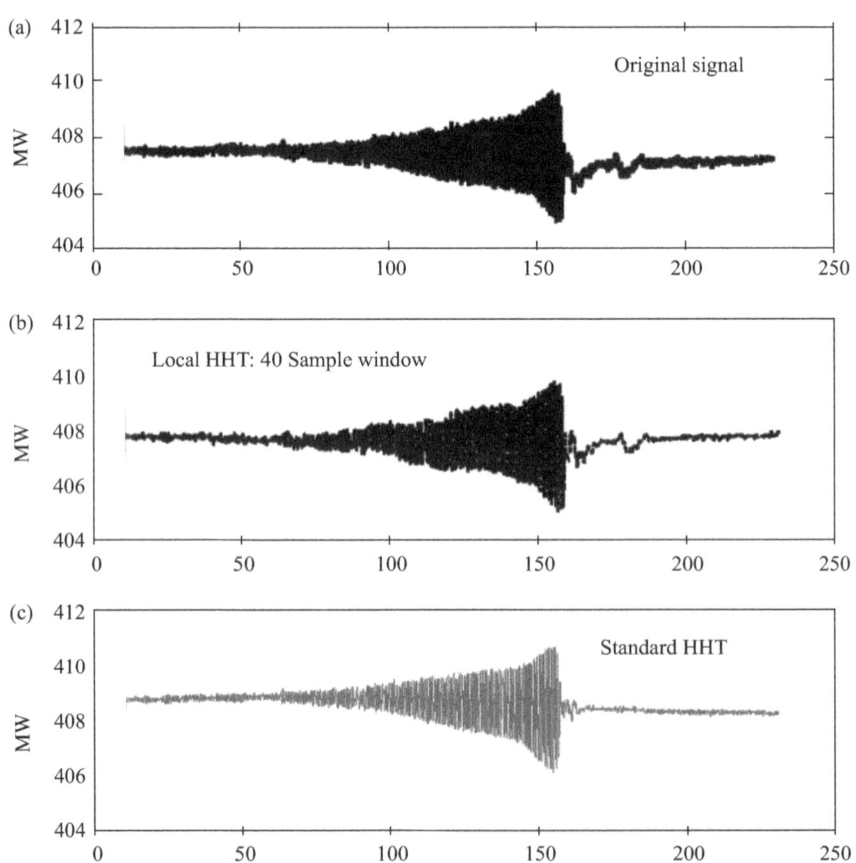

Figure 7.15 Reconstructed signals using local and block-wise implementations of HHT analysis: (a) original signal; (b) reconstructed signal using block-wise analysis; (c) reconstructed signal using conventional HHT analysis

Table 7.3 Comparison of root-mean square error for various window sizes

HHT technique	**Window size**	**MSEE**
Offline	Whole data record	0.2437
Online	60	0.0642
Online	40	0.0463
Online	30	0.0442

Table 7.4 Central processing unit (CPU) time

HHT technique	**Window size (samples)**	**CPU time (s)**
Off-line	Whole data record	0.193
Online	60	0.134
Online	40	0.122
Online	30	0.087

For a quantitative measure of the goodness of prediction of the models, the following statistics is considered

$$MSEE = \|x(t) - \hat{x}(t)\| \tag{7.23}$$

where $x(t)$ is the original signal and $\hat{x}(t)$ is the reconstructed signal from (7.5).

Table 7.3 shows the root-mean-square error for various window sizes, while Table 7.4 shows the CPU time. As shown in Table 7.3, the error decreases as the size of the sliding window in Figure 7.14 decreases. The analysis suggests that near real-time implementations of HHT analysis become more accurate.

Visual analysis of the reconstructed signal shows that the local HHT implementation captures more precisely local effects in signal's behavior.

Recent experience with measured data shows that accurate enough results may be obtained with shorter window sizes in the order of 6 samples per second [29].

7.7.3 Sliding window-based Koopman mode analysis

The developed procedures can be applied to Koopman mode analysis for transient characterization of system behavior. As pointed out in Chapter 4, Koopman mode analysis processes data using a single observation window. As a result, a single modal estimate is obtained for each time interval of interest.

The sensitivity of Koopman mode analyses to the size of the observation period has not been investigated in detail in the power system literature. This is an open issue that warrants further research.

7.8 Recursive processing methods

7.8.1 State-space model for linear regression

An interesting alternative to real-time recursive monitoring and visualization of system behavior is the use of adaptive Kalman filtering. The strength of the adaptive Kalman filter is that it provides real-time estimates of system behavior that can be post-processed with a multivariate data analysis technique.

In this section, a state-space model for linear regression with drift is used to estimate system behavior [8]. Following Sarkka [6], assume that the underlying behavior of a given signal $x(k)$ is linear and that the difference between adjacent time points is given by

$$\Delta t_{k-1}\dot{x} = x_k - x_{k-1} \tag{7.24}$$

where $\dot{x}$ denotes the derivative and $\Delta t_{k-1} = t_k - t_{k-1}$ is the time difference between consecutive times, and k represents the discrete time index.

The model can be written as a linear stochastic difference equation

$$\begin{aligned} x_k &= x_{k-1} + \Delta t_{k-1}\dot{x}_{k-1} + q_{k-1} \\ \dot{x}_k &= \dot{x}_{k-1} + \dot{q}_{k-1} \\ y_k &= x_k + r_k \end{aligned}$$

where y_k is the measurement signal, and q_k and r_k are uncorrelated zero-mean Gaussian white-noise sequences with covariance matrices $\mathbf{Q}_k$ and $\mathbf{R}_k$, respectively.

An underlying assumption is that both the noise components $r_k\ N(0,\sigma^2)$ and $(\dot{q}_k, \dot{q}_{k-1}) \sim \mathbf{N}(\mathbf{0}, \mathbf{Q})$ are independently distributed.

The model can be rewritten in a Bayesian framework assuming that the derivative performs a *random walk*:

$$\begin{aligned} p(y_k|x_k) &= N(y_k|\mathbf{H}x_{k-1}, \sigma^2) \\ p(x_k|x_{k-1}) &= N(x_k|\mathbf{A}_{k-1}x_{k-1}, \mathbf{Q}) \end{aligned} \tag{7.25}$$

where $\mathbf{Q}$ is the covariance of the random walk, $N(.)$ denotes the Gaussian probability density function, and $\mathbf{A}$ and $\mathbf{H}$ are defined as

$$\mathbf{A}_{k-1} = \begin{bmatrix} 1 & \Delta t_{k-1} \\ 0 & 1 \end{bmatrix}; \quad \mathbf{H} = [1 \quad 0]$$

The estimation of the time-series parameters proceeds through the Kalman filter; the process can be described by the following equations [30] (refer to Chapter 4):

1. Prediction equations:

$$\begin{aligned} \hat{\mathbf{x}}_{t|t-1} &= \mathbf{F}\hat{\mathbf{x}}_{t-1|t-1} \\ \mathbf{P}_{t|t-1} &= \mathbf{F}\mathbf{P}_{t-1|t-1}\mathbf{F}^T + \mathbf{G}\mathbf{Q}\mathbf{G}^T \end{aligned} \tag{7.26}$$

2. Correction equations:

$$
\begin{aligned}
\hat{\mathbf{x}}_{t|t-1} &= \mathbf{F}\hat{\mathbf{x}}_{t-1|t-1} \\
\mathbf{P}_{t|t-1} &= \mathbf{F}\mathbf{P}_{t-1|t-1}\mathbf{F}^T + \mathbf{G}\mathbf{Q}\mathbf{G}^T \\
\mathbf{\Lambda}_t &= \hat{\mathbf{x}}_{t|t-1} = \mathbf{F}\hat{\mathbf{x}}_{t-1|t-1} \\
\mathbf{P}_{t|t-1} &= \mathbf{F}\mathbf{P}_{t-1|t-1}\mathbf{F}^T + \mathbf{G}\mathbf{Q}\mathbf{G}^T \\
\mathbf{S}_t &= \mathbf{H}\mathbf{P}_{t|t-1}\mathbf{H}^T + \mathbf{R}_t \\
\mathbf{K}_t &= \mathbf{P}_{t|t-1}\mathbf{H}^T\mathbf{S}_t^{-1} \\
\hat{\mathbf{x}}_{t|t} &= \hat{\mathbf{x}}_{t|t-1} + \mathbf{K}_t\mathbf{\Lambda}_t \\
\mathbf{P}_{t|t} &= (\mathbf{I} - \mathbf{K}_t\mathbf{H})\mathbf{P}_{t|t-1}
\end{aligned} \tag{7.27}
$$

The reader is referred to [6, 8] for a detailed derivation of the model.

To illustrate these ideas consider the measured frequency signal in Figure 7.16a. Measurements are recorded over 400 s collected at a rate of 20 samples per second. Table 7.5 shows the parameters used in the Kalman filtering algorithm.

Figure 7.16b shows the result of tracking the measured signal with the Kalman filter using the above linear state-space model. For the purpose of comparison, the same signal is analyzed using the HHT method. In this analysis, HHT analysis results in seven modes and a trend. The reconstructed signal is obtained as

$$x(t) = \sum_{j=1}^{p} c_j(t) + r_k$$

Results are found to correlate very well showing the potential of these techniques for near real-time applications.

7.8.2 *Adaptive tracking of system oscillatory modes*

Recently, ambient analysis techniques to estimate power system low-frequency electromechanical modes when the primary sources of excitation are random load changes have developed [1], [3], [6]. Nonstationary RLS algorithms are especially well suited to the analysis of ambient data. A useful overview is given in [2].

In the author's previous work, a nonstationary RLS algorithm that accounts for random time-variations in the power system time series was proposed [8]. Figure 7.17 shows a conceptual representation of the identification problem using an adaptive RLS algorithm. Here, the vectors $\tilde{y}_k$ and $\hat{y}_k$ represent, respectively, the measured outputs from the power system contaminated by additive noise, v_k, which is assumed to be white noise, and the estimate of the desired (noise free) response; subscript k refers to time.

The estimation error is given by $u_k = y_k - \tilde{y}_k$ and is assumed to be white noise with variance σ_u^2. The central goal of such analysis is to track the evolving

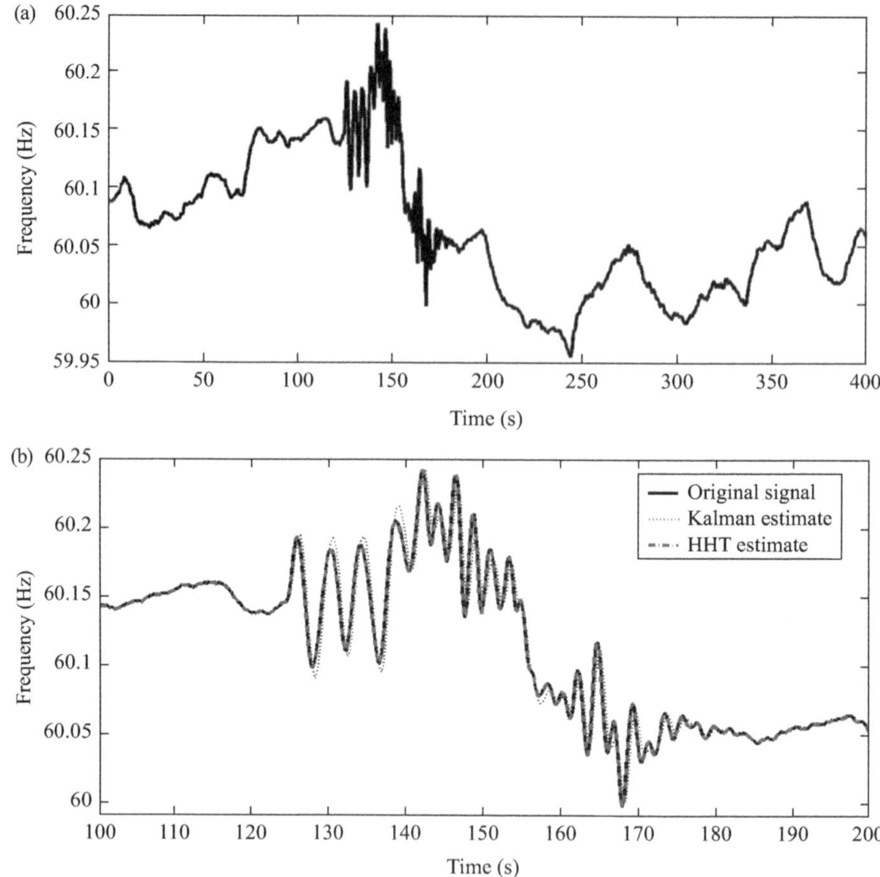

Figure 7.16 Test signal and HHT and Kalman filtering: (a) original measured signal; (b) analytic estimates using HHT and Kalman filtering

Table 7.5 Parameters of the example

Parameter description	**Numerical value**
Initial mean	$\mathbf{M} = [0 \quad 0]^T$
Initial covariance	$\mathbf{P} = diag([0.1 \quad 2])$
Measurement noise variance[a]	$\mathrm{P} = \sigma^2 = 0.1^2$
Measurement matrix	$\mathbf{H} = [1 \quad 0]$
Process noise variance[a]	$q = 0.1$
State matrix	$\mathbf{F} = \begin{bmatrix} 0 & 1 \\ 0 & 0 \end{bmatrix}$

[a]Adjustable parameters.

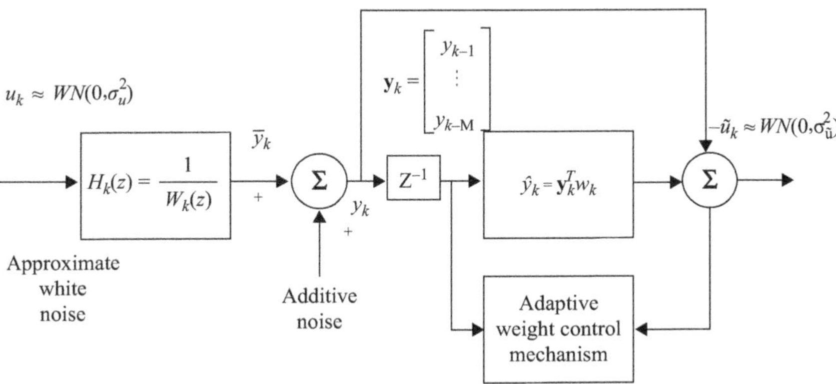

Figure 7.17 Stochastic system identification by using adaptive filtering [8]

dynamics of critical electromechanical modes present in the data, y_k, using a fully adaptive filtering technique. This problem has been addressed before in the context of RLS adaptive filtering techniques [1, 3, 8].

In developing the algorithm, consider a stochastic process that admits a state-space representation of the form

$$\begin{aligned} \mathbf{x}_{k+1} &= \lambda^{-1/2}\mathbf{x}_k + \mathbf{q}_k \\ y_k &= \mathbf{c}_k\mathbf{x}_k + r_k \end{aligned} \tag{7.28}$$

where λ is a forgetting factors, $\mathbf{x}_k$ is the state vector of the stochastic process model at time k, $\mathbf{c}_k$ is known $1 \times M$ vector, $\mathbf{q}_k$ is a process (state) noise vector, y_k is the observation signal, and r_k is the additive measurement noise. It is assumed that $\mathbf{q}_k$ and r_k are uncorrelated zero-mean stochastic vectors with correlation matrices $\mathbf{Q}_k$ and $\mathbf{R}_k$, respectively.

Under some simplifications the model (7.28) [6] transforms to $\mathbf{x}_k = \lambda^{-1/2}\mathbf{x}_k$, $y_k = \mathbf{y}_k^T\mathbf{x}_k + r_k$ (see [31] for details). Some implications of this model are worth emphasizing. First, the process noise is assumed to be zero. As a consequence, the RLS mechanism cannot adapt itself to fast changes in signal characteristics during transients and is therefore not suited for nonstationary environments.

In [8], a nonstationary RLS algorithm that accounts for random time-variations in the measured power system time series that circumvents the above limitations was developed. The algorithm is designed by assuming that a random walk process describes the state evolution in x_k over time.

In this approach, the conventional model (7.28) is rewritten in the form

$$\begin{aligned} \mathbf{x}_k &= \lambda^{-1/2}\mathbf{x}_k + \Delta\mathbf{x}_k \\ \mathbf{y}_k &= \mathbf{y}_k^T\mathbf{x}_k + r_k \end{aligned} \tag{7.29}$$

in which the random variables are modeled by the term $\Delta\mathbf{x}_k$ which has the same statistical characteristics as $\mathbf{q}_k$.

Application of the Kalman filter to the above model results in a recursive set of equations in which the forgetting factor is determined recursively. The application of the Kalman filter theory to the model (7.28) yields the recursive equations [6]

$$\begin{cases} \tilde{\mathbf{x}}_k = \lambda^{-1/2}\tilde{\mathbf{x}}_{k-1} \\ \tilde{\mathbf{P}}_k = \lambda^{-1}\tilde{\mathbf{P}}_{k-1} + \mathbf{Q}_{k-1} \\ \hat{\mathbf{x}}_k = \tilde{\mathbf{x}}_k + \tilde{\mathbf{P}}_k\mathbf{c}_k^T\left(\mathbf{c}_k\tilde{\mathbf{P}}_k\mathbf{c}_k^T + \mathbf{R}_k\right)^{-1}\left(y_k - \mathbf{c}_k\tilde{\mathbf{x}}_k\right) \\ \hat{\mathbf{P}}_k = \tilde{\mathbf{P}}_k - \tilde{\mathbf{P}}_k\mathbf{c}_k^T\left(\mathbf{c}_k\tilde{\mathbf{P}}_k\mathbf{c}_k^T + \mathbf{R}_k\right)\mathbf{c}_k\tilde{\mathbf{P}}_k \end{cases} \tag{7.30}$$

where $\hat{\mathbf{x}}_k$ is the linear least-mean squares solution for the model in (7.10), and $\hat{\mathbf{P}}_k$ is the covariance matrix of the state-estimation error, with initial conditions

$$\hat{\mathbf{x}}_o = \lambda_k\mathbf{B}_M^{-1}\mathbf{y}_M$$
$$\hat{\mathbf{P}}_o = Var(\hat{\mathbf{x}}_o - \mathbf{x}_o) = Var(\mathbf{v}_M - \lambda_k\mathbf{B}_M^{-1}\mathbf{w}_{1|M})$$

This approach improves the tracking capability of the method to deal with abrupt changes in ambient data. The application of nonlinear adaptive recursive least-squares (NRLS) techniques requires the solution of three main problems:

1. The computation of the variable forgetting factor
2. The determination of empirical estimators for the noise statistics
3. The computation of initial conditions

Refer to [8] for specific details about the implementation of the method.

7.8.2.1 Damping estimation

Kalman filtering is especially well suited for damping identification in real time. A simple alternative is to represent the observed behavior by a second order model of the form [32].

Let the measured data, y_k, be expressed in the form

$$y_k = \sum_{j=1}^{L} A_j e^{(-\sigma_j + 2\pi f_j)kT_s} \tag{7.31}$$

$k = 1, \ldots, N$, where A_j, σ_j and f_j are the modal parameters and T_s is the sampling period.

Computation of modal parameters involves two steps:

1. Obtain a discrete state-space representation of the model as
$$\begin{aligned} \mathbf{x}_{k+1} &= \mathbf{f}(\mathbf{x}_k) \\ y_k &= \mathbf{H}\mathbf{x}_k + v_k \end{aligned} \tag{7.32}$$
2. Compute modal parameters in (7.32) with
$$x_k^1 = e^{(-\sigma_j + 2\pi f_j)kT_s}$$
$$x_k^2 = A_j e^{(-\sigma_j + 2\pi f_j)kT_s}$$

where the terms in these expressions have the usual interpretations.

Other more general interpretations can be obtained directly from a second-order representation of a second-order degree-of-freedom (SDOF) oscillator.

The nonstationary RLS algorithm can be summarized as follows – See [9] for details and equivalences with the Kalman variables and intermediate steps.

Nonstationary RLS algorithm with variable forgetting vector

Given a set of simultaneously recorded signals $x_k(t)$, $k = 1, \ldots, m$. For each time instance k, $k = 1, \ldots, m$

1. Determine the initial conditions $\hat{\mathbf{x}}_o, \hat{\mathbf{P}}_o, \Psi_o, \lambda_o$. Set $k = 0$.
2. Set $k = k + 1$. Determine the propagation of the state vector from

$$\tilde{\mathbf{x}}_k = \lambda^{-1/2}\tilde{\mathbf{x}}_{k-1}$$
$$\tilde{\mathbf{P}}_k = \lambda^{-1}\tilde{\mathbf{P}}_{k-1} + \mathbf{Q}_{k-1}$$

3. Given initial conditions, compute the gain vector and *a priori* error $\upsilon_k, \kappa_k, e_k$ using

$$\overline{\mathbf{P}}_k = \lambda_k^{-1}\mathbf{R}_{k-1}\tilde{\mathbf{P}}_{k-1}\mathbf{Q}_{k-1}$$
$$\mathbf{\upsilon}_k^{-1} = \left(\mathbf{y}_k^T\overline{\mathbf{P}}_k\mathbf{y}_k\mathbf{R}_k\right)^{-1}$$
$$\kappa_k = \overline{\mathbf{P}}_k\mathbf{y}_k\mathbf{\upsilon}_k^{-1}$$

4. Compute the weight vector

$$\mathbf{w}_k = \mathbf{w}_{k+1} + \boldsymbol{\kappa}_k e_k$$

   Compute modal properties (frequency and damping ratio) at each time instant k.
5. Compute the correlation matrix inverse $\mathbf{P}_k$.
6. If $k = N$, return to step 2.

Table 7.6 compares the NRLS algorithm with other nonlinear and/or nonstationary methods.

Relevant potential applications of this model to measured data include the following:

- Extraction of time-varying trends
- Identification of modal parameters, namely damping, frequency, and amplitude information

Table 7.6 Parameters of the example

Technique	Realtime	Mean extraction	Frequency/ Damping	Phase	Observations
NRLS	Yes	Yes	Yes	No	Under development
DHR	No	Yes	No	No	Under development
HHT	Yes	Yes	Yes	Yes	Near real-time

Since the method is nonstationary, it is ideally suited to incorporate multisensory multitemporal data fusion techniques. This is a subject of future research.

References

1. Daniel J. Trudnowski, John W. Pierre, Ning Zhou, John F. Hauer, Manu Parashar, 'Performance of three mode-meter block-processing algorithms for automated dynamic stability assessment', *IEEE Transactions on Power Systems*, vol. 23, no. 2, May 2008, pp. 680–690.
2. Power System Dynamic Performance Committee, Task Force on Identification of Electromechanical Modes, Chair: Juan J. Sánchez Gasca, 'Identification of electromechanical modes in power systems', IEEE/PES Special Publication TP462, June 2012.
3. N. Zhou, J. W. Pierre, R. W. Wies, 'Estimation of low-frequency electromechanical modes of power systems from ambient measurements using a subspace method', 2003 North American Power Symposium, 2003.
4. Arturo R. Messina (ed.), *Inter-area Oscillations in Power Systems – A Nonlinear and Nonstationary Perspective*, Power Electronics and Power Systems Series, Springer Science, New York, NY, 2009.
5. N. Zhou, J. Pierre, J. F. Hauer, 'Initial results in power system identification from injected probing signals using a subspace method', *IEEE Transactions on Power Systems*, vol. 21, no. 3, August 2006, pp. 1296–1302.
6. Simmo Särkkä, *Bayesian Filtering and Smoothing*, Cambridge University Press, New York, NY, 2013.
7. N. Zhou, J. W. Pierre, D. J. Trudnowski, R. T. Guttromson, 'Robust RLS methods for online estimation of power system electromechanical modes', *IEEE Transactions on Power Systems*, vol. 22, no. 3, August 2007, pp. 1240–1249.
8. I. Moreno, A. R. Messina, 'Adaptive tracking of system oscillatory modes using an extended RLS algorithm', *Electric Power Systems Research*, vol. 114, 2014, pp. 28–38.
9. P. Korba, 'Real-time monitoring of electromechanical oscillations in power systems: First findings', *IET Generation, Transmission & Distribution*, vol. 1, no. 1, January 2007, pp. 80–88.
10. Mladen Kezunovic, Sakis Meliopoulos, Vaithianathan Venkatasubramanian, Vijay Vittal, *Applications of Time-Synchronized Measurements in Power Transmission Networks*, Power Electronics and Power Systems Series, Springer, Cham, Switzerland, 2014.
11. Arturo R. Messina, Vijay Vittal, Gerald T. Heydt, Timothy J. Brown, 'Nonstationary approaches to trend identification and denoising of measured power system oscillations', *IEEE Transactions on Power Systems*, vol. 24, no. 2, November 2009, pp. 1798–1806.
12. Alexander Bykhovsky, Joe H. Chow, 'Power system disturbance identification from recorded dynamic data at the Northfield substation', *Electrical Power and Energy Systems*, vol. 25, 2003, pp. 787–795.

13. David B. Bertagnolli, Xiachouan Luo, James W. Ingleson, Joe H. Chow, J. Gregory Allcorn, Mark Kuras, Harish I. Mehta, ... James P. Hackett, 'Northeastern US oscillation detection and recording project', Fault and Disturbance Analysis Conference, April 2004.
14. David L. Donoho, Iain M. Johnstone, 'Ideal spatial adaptation by wavelet shrinkage', *Biometrika*, vol. 81, no. 3, 1994, p. 425.
15. Mina Aminghafari, Natalie Cheze, Jean-Michel Poggi, 'Multivariate denoising using wavelets and principal component analysis', *Computational Statistics & Data Analysis*, vol. 50, 2006, pp. 2381–2398.
16. J. F. Hauer, F. Vakili, 'An oscillation detector used in the BPA power system disturbance monitor', *IEEE Transactions on Power Systems*, vol. 5, 1990, pp. 74–79.
17. A. Bykhovsky, Joe Chow, 'Power system disturbance identification from recorded dynamic data at the Northfield substation', *International Journal of Electrical Power & Energy Systems*, vol. 25, no. 10, 2003, pp. 787–795.
18. Richard P. Schulz, Beverly B. Laios, 'Triggering tradeoffs for recording dynamics', *IEEE Computer Applications in Power*, April 1997, pp. 44–49.
19. Szi-Wen Chen, Hsiao-Chen-Chen, Hsiao-Lung Chan, 'A real-time QRS detection method based on moving average incorporating with wavelet denoising', *Computer Methods and Programs in Medicine*, vol. 82, 2006, pp. 187–195.
20. Desiree Phillips, Thomas Overbye, 'Distribution system event detection and classification using local voltage measurements', 2014 Power and Energy Conference, Illinois.
21. Penn Markham, Ye Zhang, Yilu Liu, John Stovall, Marcus Young, Jose Gracia, Thomas King, 'Wide-area power system frequency measurement applications', *Future of Instrumentation International Workshop (FIIW), 2012.* Gatlinburg, TN, 8–9 October 2012. IEEE, 2012. http://ieeexplore.ieee.org/xpl/mostRecentIssue.jsp?punumber=6362372
22. A. R. Messina, Vijay Vittal, Daniel Ruiz-Vega, G. Enríquez Harper, 'Interpretation and visualization of wide-area PMU measurements using Hilbert analysis', *IEEE Transactions on Power Systems*, vol. 21, no. 4, November 2006, pp. 1763–1771.
23. F. L. Zarraga, A. L. Rios, P. Esquivel, A. R. Messina, 'A Hilbert–Huang based approach for online extraction of modal behavior from PMU data', 2009 North American Power Symposium.
24. Patrick Flandrin, Paulo Goncalves, 'Empirical mode decompositions as data-driven wavelet-like expansions', *International Journal of Wavelets, Multiresolution and Information Processing*, vol. 2, no. 4, 2004, pp. 477–496.
25. R. L. C. Spaendonck, F. C. A. Fernandes, R. G. Baraniuk, and J. T. Fokkema, 'Local Hilbert transformation for seismic attributes', *Proceedings of the EAGE 64th Conference and Exhibition*, Florence, Italy, May 2002.
26. Dina S. Laila, Arturo R. Messina, Bikash C. Pal, 'A refined Hilbert–Huang transform with application to interarea oscillation monitoring', *IEEE Transactions on Power Systems*, vol. 24, no. 2, May 2009, pp. 610–620.

27. Arthur E. Barnes, 'The calculation of instantaneous frequency and instantaneous bandwidth', *Geophysics*, vol. 57, no. 11, November 1992, pp. 1520–1524.
28. M. T. Taner, F. Koehler, R. E. Sheriff, 'Complex seismic trace analysis: Geophysics', vol. 44, 1979, pp. 1041–1063.
29. F. Lezama, *Use of the Hilbert Transform to Analyze Synchrophasor Data*, PhD Thesis, The Center for Research and Advanced Studies, Cinvestav, Guadalajara, Mexico, 2011.
30. R. Khon, C. F. Ansley, 'Filtering and smoothing algorithms for state space models', *Computers & Mathematics with Applications*, vol. 18, nos. 6/7, 1989, pp. 515–528.
31. D. Godard, 'Channel equalization using a Kalman filter for fast data transmission', *IBM Journal of Research and Development*, vol. 18, no. 3, May 1974, pp. 283–300.
32. Jimmy C. -H. Peng, Nirmal-Kumar C. Nair, Jian Zhang and Akshya Kumar Swain, Detection of Lightly Damped Inter-Area Power Oscillations using Extended Complex Kalman Filter, 2009 IEEE Region 10 Conference (TENCON 2009), Singapore, January 2009.

Chapter 8
Interpretation and visualization of wide-area PMU measurements

8.1 Introduction

The analysis of multiple sets of data usually of different type or nature is challenging problem in power system stability analysis. Examination of system disturbances may involve a large number of measured signals with composite record lengths on the order of several minutes or hours [1–3] and be complicated by noise, trends and other artifacts.

In addition, comparisons are also needed against models simulations, dynamic probing tests, and previous events [4, 5]. Records collected on the wide-area monitoring systems (WAMS) are contaminated by noise from different sources [6, 7].

In this chapter, measured data from an actual system event are used to investigate the ability of wide-area monitoring techniques to monitor and visualize system behavior. Several multi-sensor data fusion-based forecasting architectures are investigated and tested. The applications covered include the assessment, and use of various signal processing techniques to measured synchrophasor data.

Practical methods for obtaining approximations to system behavior are discussed and the accuracy of the models is evaluated. Visualization techniques are also presented.

The experience in the analysis of collected data from phasor measurement units (PMUs) is discussed. The issues of data collection, conditioning, and extraction of the primary oscillation frequency are discussed.

8.2 Loss of generation oscillation event

Synchronized phasor measurements of a real event [8] in the Mexican Interconnected System (MIS) are used to test the ability of multisensor data fusion techniques to detect and localize damage in the presence of abnormal system conditions in a large power system.

The data set comprises time series of key system parameters recorded at 18 separate locations across the MIS using PMUs. Each time history is recorded for about 250 s and consists of 4 900 data points.

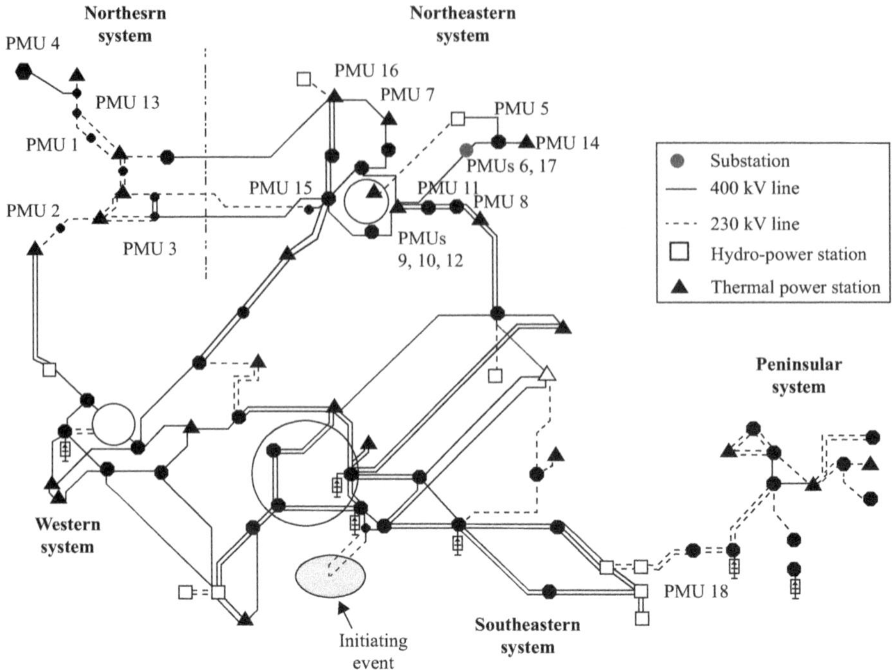

Figure 8.1 Schematic of the MIS showing the location of installed PMUs and the monitored areas for wide-area monitoring. The location of the initiating event is shown as the hatched region of the plot

A geographical diagram of the system showing portions of interest of the MIS and the location of the monitoring sites as well as the main areas or regions of interest are shown in Figure 8.1. Section 8.2.1 summarizes the context in which these oscillations occurred.

8.2.1 Operational context

On July 4, 2004, oscillations involving power frequency and voltage were observed at the MIS. The main event that originated the oscillations was an increase in generating power in a remote hydroelectric generating station in the southwestern system, followed by a 600 MW generation loss. This station consists of three 200 MW identical generators connected to the 230 kV transmission network through a two-circuit transmission line as shown schematically in Figure 8.2.

Previous to this event, circuit #1 of the two-circuit 230 kV transmission line connecting this generator to the system was out of service [8]. This excited a local electromechanical mode involving the interaction of this plant with the rest of the system.

8.2.2 Recorded measurements

Undamped system oscillations developed for about 170 s until the second circuit was tripped by overloading, resulting in generation rejection (see Figure 8.2). As a

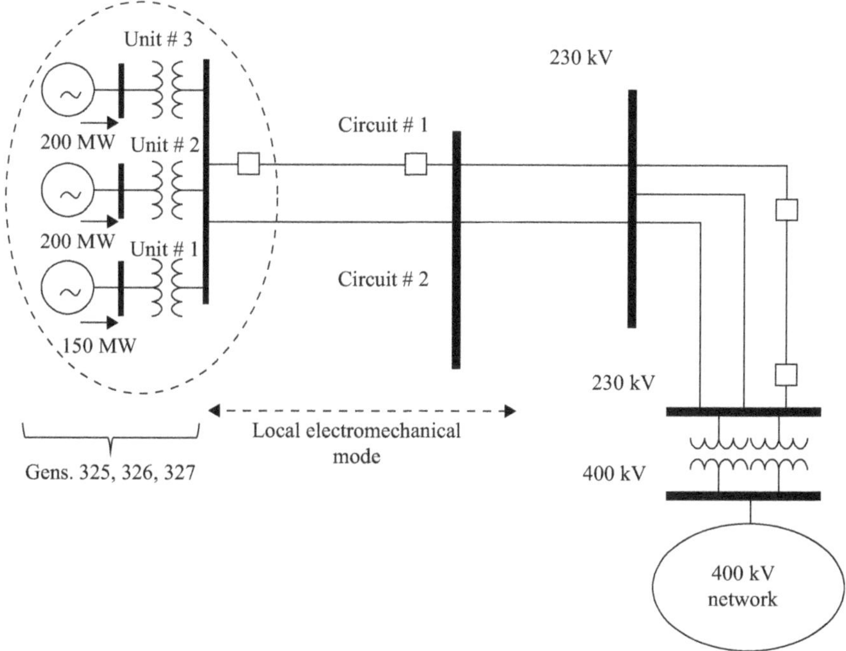

Figure 8.2 Schematic of the local hydroelectric generating station: pre-disturbance operating condition associated with a local electromechanical mode. Empty squares denote transmission circuits out of service

Table 8.1 Summary of system events

1. At the time of the July 31 event, the power output of a remote hydroelectric plant in the southeastern network of the system was 542 MW with units 2 and 3 online. Units 2 and 3 were operating at 196 MW. Unit 3 was operating at 150 MW.
2. System oscillations then developed when the power output of Unit 3 was increased from 150 MW to 196 MW.
3. At 14:45:51.900, circuit #2 tripped open due to excessive overloading; this event isolated 588 MW of generation from the rest of the system.
4. The frequency drops to about 59.74 Hz to then increase slowly to a new steady-state value of about 60 Hz.

result, the frequency dropped at about 59.7 Hz due to the tripping of about 600 MW to then recover slowly to 60 Hz. A summary of the sequence of events leading to the observed undamped electromechanical oscillations is shown in Table 8.1 [8].

Among the existing network of PMUs, measurements from three regional systems are selected for analysis; the type, sampling rate, and locations of sensors are given in Table 8.2. Geographically, measurements included three regions, the north-western portion of the system, the northern portion, and the southeastern portion.

Table 8.2 PMU measurement locations

PMU number	Area	Sampling rate (sps)
1, 2, 3, 4, 13	Northern system	20
5, 6, 7, 8, 9, 10, 11, 12, 14, 15, 16, 17	Northeastern system	20
18	Southeastern system	20

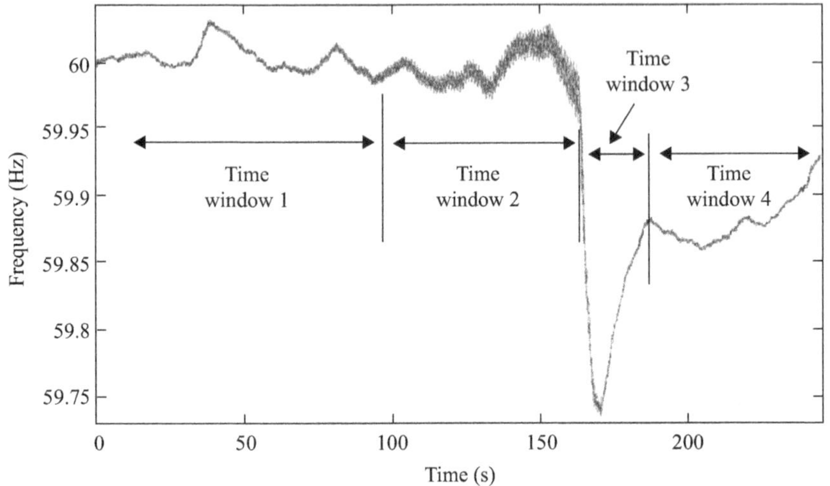

Figure 8.3 Frequency transients for the generator trip event

Selected frequency, voltage, and power flow recordings at major substation from selected PMUs are shown in Figure 8.3 through 8.5, presenting time intervals of interest in the study. As shown, oscillations can be detected in all the recorded variables. Worthy of interest, a strong trend can be noted in the recordings associated with various control and switching actions. Inter-area oscillations are also evident associated with the exchange of energy in the system as discussed below.

For purposes of illustration the data was divided into four segments or time windows of particular interest:

1. *Time window 1 (0–100 s):* This is the time window subsequent to the start of the measurements and is dominated by ambient behavior.
2. *Time window 2 (100–164 s):* During this interval system motion is dominated by a local mode at about 1.0 Hz.
3. *Time window 3 (164–187 s):* This time interval includes the inertial response and the start of automatic generation control (AGC) action.
4. *Time window 4 (187–250 s):* Time interval 4 starts when the AGC function ends and shows an oscillation in which an inter-area mode at about 0.4 Hz is visible.

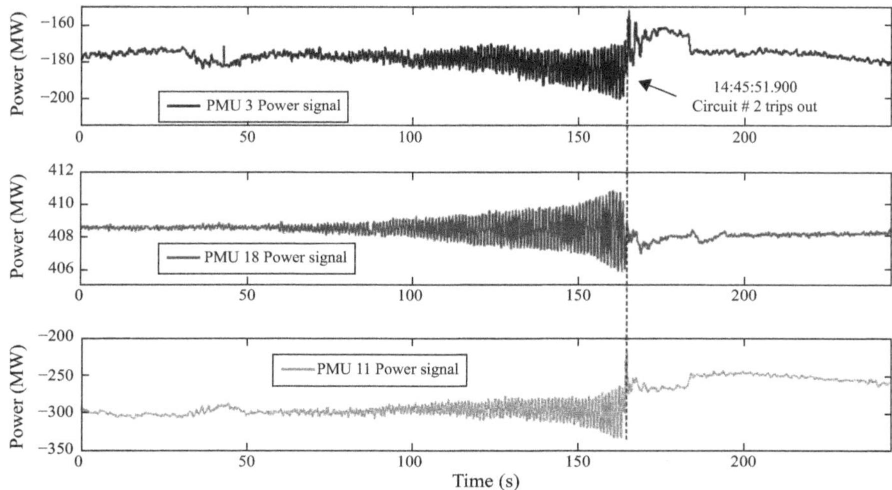

Figure 8.4 Selected time traces of recorded power flows

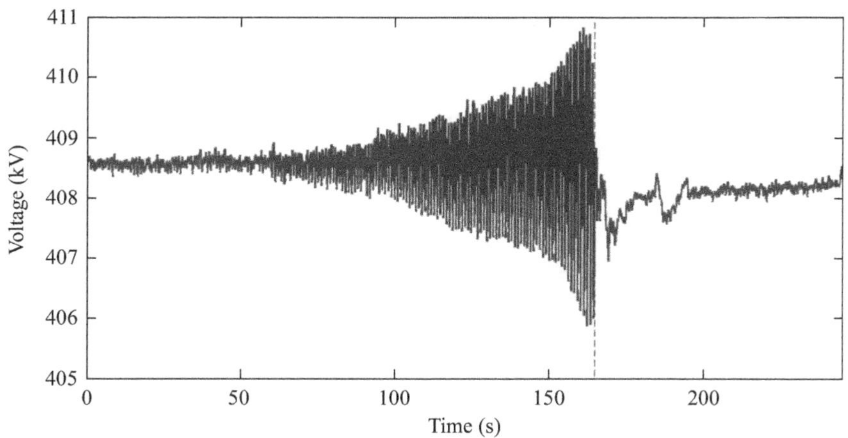

Figure 8.5 Recorded 400 kV bus voltage magnitude at PMU 13

Spectral analysis of the system behavior shown in Figure 8.6 indicates a mode near 1.00 Hz that is strongly observable mainly at PMUs 18 and 4 at the end of the southeastern and northern systems where the frequencies experience the most fluctuation. A second mode near 0.47 Hz is also observable associated with measurements at PMUs 5, 11, and 15 [9]. The relatively smaller peak at about 0.52 Hz is associated with the interaction between buses in the northern systems (PMUs 1 and 2) and buses in the northeastern system (PMUs 12 and 18).

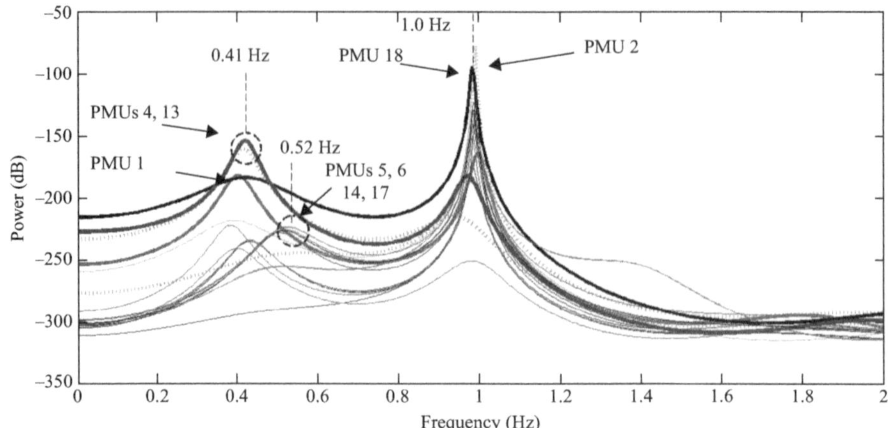

Figure 8.6 AR spectra of frequency measurements in Figure 8.3

8.3 Analysis and visualization of recorded data

Measured data from PMUs was collected and used for global system analysis. Two main WAMS strategies were considered in the analysis of system measurements: (a) a fully centralized architecture and (b) a decentralized architecture.

In the first case, the m-by-N matrix of observations $\mathbf{X}$ was created from the snapshots of raw measurements as

$$\mathbf{X} = [\mathbf{x}_1 \quad \mathbf{x}_2 \quad \ldots \quad \mathbf{x}_m] \tag{8.1}$$

where m is the number of *signals*, $\mathbf{x}_j$, and N is the number of time points in a given interval. For reference, spectral analyses were conducted for all measured signals.

In the second case, analyses were conducted using the actual geographical location of power data concentrators in the system [10].

The analysis is focused on three aspects, namely, mode shape estimation, damping calculation, and the analysis of temporal behavior. These aspects are discussed separately.

8.3.1 Mode shape characterization

As discussed earlier, the mode shapes can be interpreted as the spatial energy distributions associated with the oscillation modes, and can be obtained from the eigenvectors of the response matrix.

In the analysis that follows, three analytical methods to estimate mode shapes from multiple synchrophasors were considered and compared:

1. Proper orthogonal decomposition/PCA
2. Blind source separation (BSS)
3. Koopman mode analysis (KMA)

Frequency measurements were initially chosen for analysis, since frequency is a global quantity [11, 12]. The frequency-based observation matrix, **X**, corresponds to equation (8.1), with m=18, that is

$$\mathbf{X} = \left[\mathbf{f}_{PMU_1} \quad \mathbf{f}_{PMU_2} \quad \ldots \quad \mathbf{f}_{PMU_m} \right]^T \tag{8.2}$$

where $\mathbf{f}_{PMU_k} = [f_{PMU_k}(t_1) \quad f_{PMU_k}(t_2) \quad \ldots \quad f_{PMU_k}(t_N)]^T$, $k = 1, \ldots, 18$.

For the PCA and BSS approaches, the ensemble of measurements was decomposed into a set of uncorrelated modal components and the mode shapes were extracted using the procedures set out in Chapters 3 and 6, respectively.

Figure 8.7 shows the extracted mode shapes for the three representations above. The analysis of the mode shape for the dominant mode in Figure 8.7 indicates three main oscillation clusters associated with machines in the northern, northeastern, and southeastern systems in Figure 8.1; these clusters correspond to the inter-area oscillations between geographical regions in Figure 8.1. For the 1.0 Hz mode, the frequency signals in the northern system (PMUs 2, 3, 4, 11, and 13) and in the southeastern system (PMU 18) are found to swing against mainly signals in the northwestern system (PMUs 1, 5–7, 9–10, 12, and 14–17). The signals at PMUs 2, 8, and 18 are found to have the strongest participation in this mode.

Results are found to be in good qualitative agreement although some differences are noted.

Section 8.3.2 outlines and compares two approaches to estimate modal damping from multivariate data, the multisignal Prony analysis method, and the Koopman mode decomposition approach.

8.3.2 Damping estimation

Two different approaches to global modal damping estimation are outlined and compared: (1) multisignal Prony (MSP) analysis based on the Tufts–Kumaresan algorithm [13] and (2) Koopman mode analysis [14]. For clarity of illustration, the accuracy and robustness of the modeling approaches were evaluated for three analysis intervals: 0–100 s and 100–164 s.

Table 8.3 compares the modal frequency and damping of the MSP method with the corresponding modal estimates of the Koopman method. Results are found to be in good agreement for the time interval 120–160s; no physically meaningful estimates were obtained using Prony analysis for the time interval 0–120 s.

8.3.3 Instantaneous parameters

To further visualize the phenomenon of mode propagation along the system, the evolutionary behavior of measured signals is examined in the time–frequency domain using both the Hilbert–Huang transform (HHT) method and wavelet analysis.

In the studies that follow, each raw measurement vector, $\mathbf{x}_j$, was decomposed in the form

$$x(t) = \sum_{j=1}^{p} c_j(t) + r_p(t) \tag{8.3}$$

where the c_j are the oscillatory components of concern.

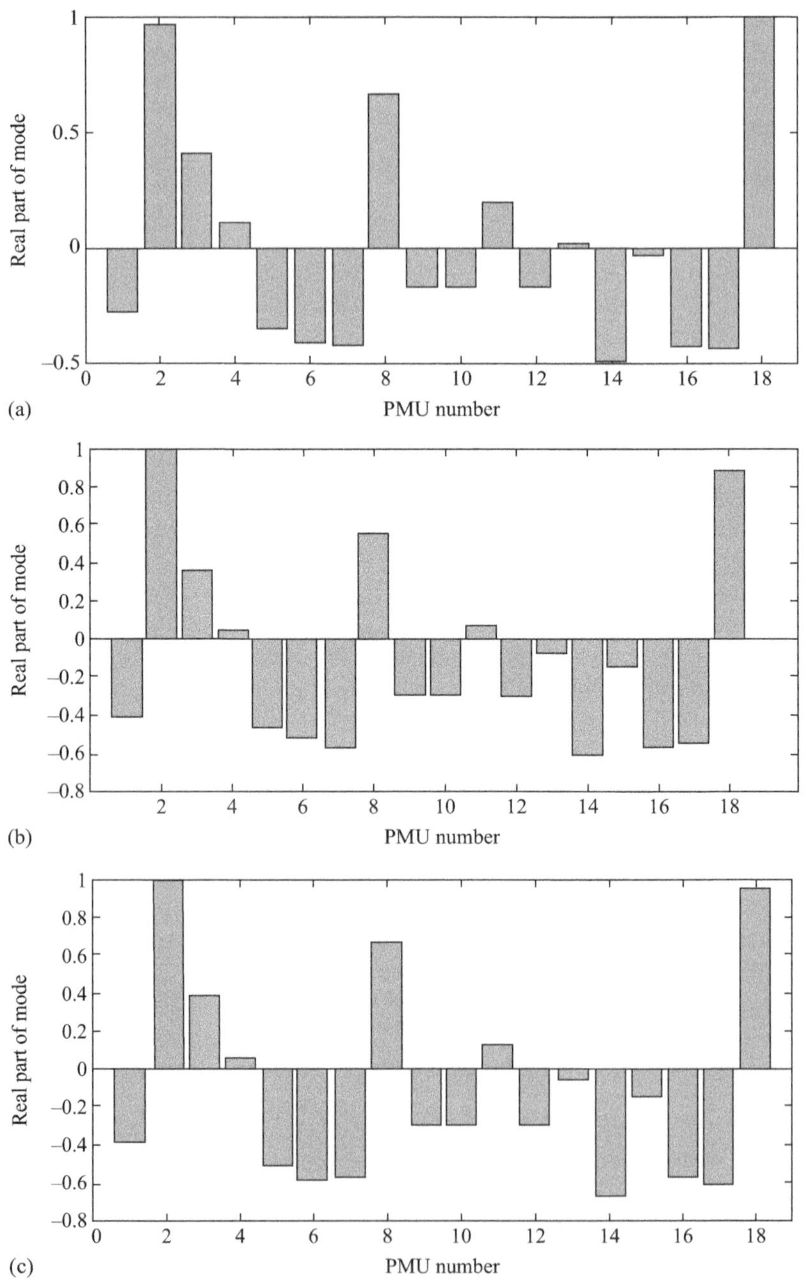

Figure 8.7 Comparison of mode shapes for the 1.0 Hz mode: (a) PCA; (b) blind source separation; (c) Koopman mode decomposition

Table 8.3 Global damping estimation using Koopman and multisignal Prony analysis. Frequency signals

Time interval	MSP analysis		Koopman mode analysis	
	Frequency (Hz)	**Damping ($\sigma/2\pi$)**	**Frequency (Hz)**	**Damping ($\sigma/2\pi$)**
0–160 s	*	*	0.986	−0.017
120–160 s	0.989	−0.018	1.084	−0.034

*No physically meaningful solution was obtained.

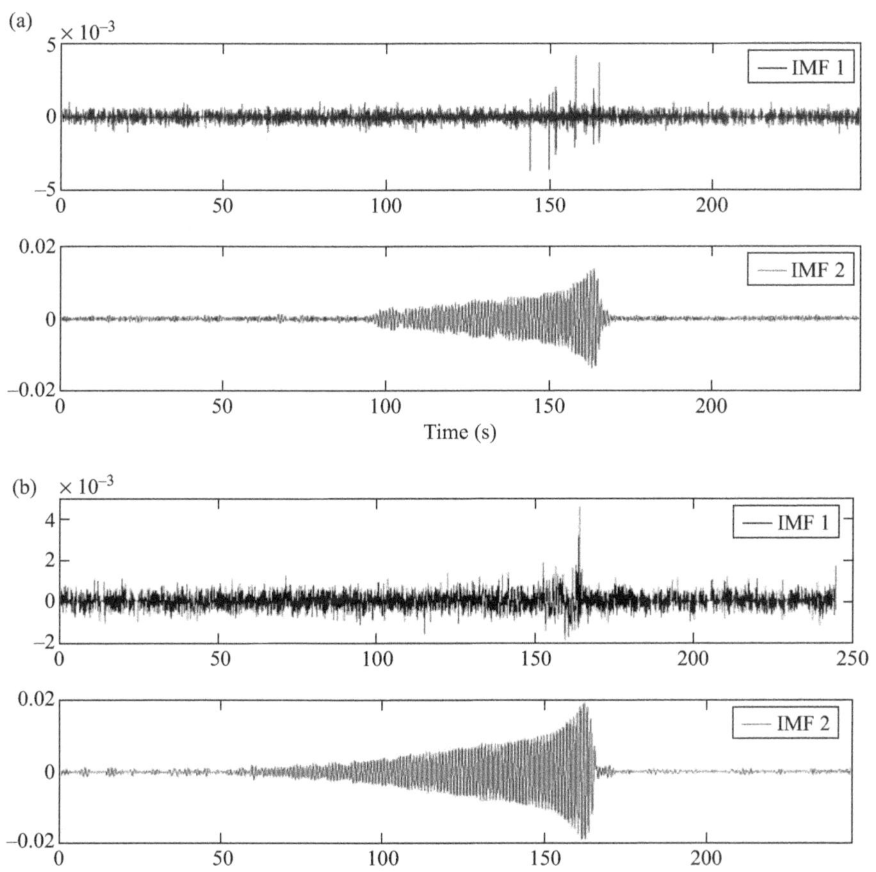

Figure 8.8 Time evolution of the first two modal components: (a) PMU 2; (b) PMU 18

For clarity of presentation, the frequency signals from PMUs 4 and 13 were selected for analysis. Figure 8.8 depicts the two leading intrinsic mode functions (IMFs) generated from the frequency signals for PMUs 2 and 18. Other IMFs make a negligible contribution to the observed response and are not considered here.

For both cases, IMF 1 is seen to capture high-frequency noise, while IMF 2 captures the temporal behavior of interest. Similar results are obtained using wavelet analysis and are presented here.

8.3.3.1 Instantaneous frequency

The nature of system behavior becomes clear in Figure 8.9 that shows the Hilbert amplitude spectrum for the frequency signals associated with IMF 2 for PMU 18. Similar results are obtained with the wavelet method.

In these studies, the instantaneous frequency was computed as

$$f(t) = \left(\frac{1}{2\pi}\right)\frac{c(t)\dot{c}_H(t) - c_H(t)\dot{c}(t)}{c^2(t) + c_H^2(t)}$$

where $c_H(t)$ is the Hilbert transform of $c(t)$.

Inspection of the amplitude spectrum in this plot for PMU 18 shows a nearly constant frequency mode centered at about 1.0 Hz; the varying shade of the contour plot suggests some degree of frequency modulation. The amplitude of these oscillations, in turn, increases with time revealing unstable behavior.

These results are consistent with the observed behavior in Figure 8.8b. Similar representations are obtained for other buses.

8.3.3.2 Instantaneous energy

Results of the previous section motivate us to develop global analysis techniques. To further analyze the nature of the energy propagation phenomena, the time evolutions of the instantaneous amplitudes for the 18 modal frequencies for the 1.0 Hz components were analyzed simultaneously using the multiscale temporal fusion approach in Chapter 5.

Based on actual information from the local PDCs, a multiblock analysis technique was applied to extract modal information from the three local PDCs. Attention was focused on the analysis of major electromechanical modes. The measured signals were locally analyzed to extract modal characteristics.

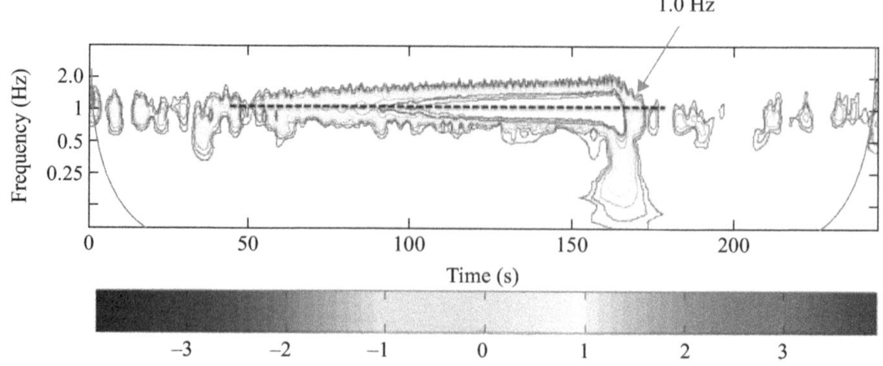

Figure 8.9 Hilbert spectrum of measured data: PMU 18

Two main analysis strategies were tested:

1. *Sensor-level fusion:* In this analysis, 18 observation matrices were constructed of the form

$$\mathbf{X}(i,t) = \begin{bmatrix} c_{i1}(t_1) & c_{i1}(t_2) & \cdots & c_{i1}(t_N) \\ c_{i2}(t_2) & c_{i2}(t_2) & \cdots & c_{i2}(t_2) \end{bmatrix}, \quad i = 1,\ldots,18 \tag{8.4}$$

for the two modes of interest at 0.42 and 1.0 Hz ($i = 1,\ 2$). Note that the superscript k has been dropped for convenience.

The corresponding feature matrix can be written as

$$\mathbf{X}_f = [\mathbf{X}(1,t) \mid \mathbf{X}(2,t) \mid \cdots \mid \mathbf{X}(18,t)]$$

This case corresponds to the hybrid multiblock PCA (POD) analysis shown in Figure 5.10.

2. *Area-level fusion:* In this case, data was fused at a PDC level (PDC 1 = northern system, PDC 2 = northeastern system, PDC 3 = southeastern system). Attention was restricted to the analysis of the local mode around 1.0 Hz.

Figure 8.10 is an illustration of the extracted spatio-temporal patterns obtained using the first approach above. Figure 8.10a shows the instantaneous amplitudes $A_j(t)$, $j = 1,\ldots,18$ calculated using Hilbert analysis. Figure 8.10b shows a three-dimensional visualization of the evolving dynamic process. For ease of comparison, identical energy scales are used for all plotted records.

The analysis suggests that instantaneous energies (amplitudes) can be used to estimate the propagation rate of inter-area phenomena as well as to coordinate control actions in real time.

When compared with the individual time–energy representations in Figure 8.9, it is apparent that the proposed approach gives a better representation for the overall system dynamics. This gives additional information regarding the strength and distribution of the propagating phenomenon which is complementary to that obtained from local analysis.

The model can be used to determine boundaries between regions or geographical zones as well as to study the frequency disturbance propagation distance and speed. Several measures of spatio-temporal dynamics can be computed such as the spatial amplitude and phase functions and the temporal amplitude and phase functions [13].

Figure 8.11 shows a similar analysis from approach 2 above. Results are found to be consistent showing the potential of the analysis for wide-area monitoring. Other formulations are also possible and are the subject of current interest.

8.3.3.3 Instantaneous damping

Figure 8.12 shows the instantaneous damping and amplitude associated with the 1.0 Hz component in Figure 8.10. For a direct comparison to the observed oscillation, the time evolution of the IMFs and instantaneous amplitude for the 0.41 Hz (IMF 3) and 1.0 Hz (IMF 2) are plotted in Figure 8.13.

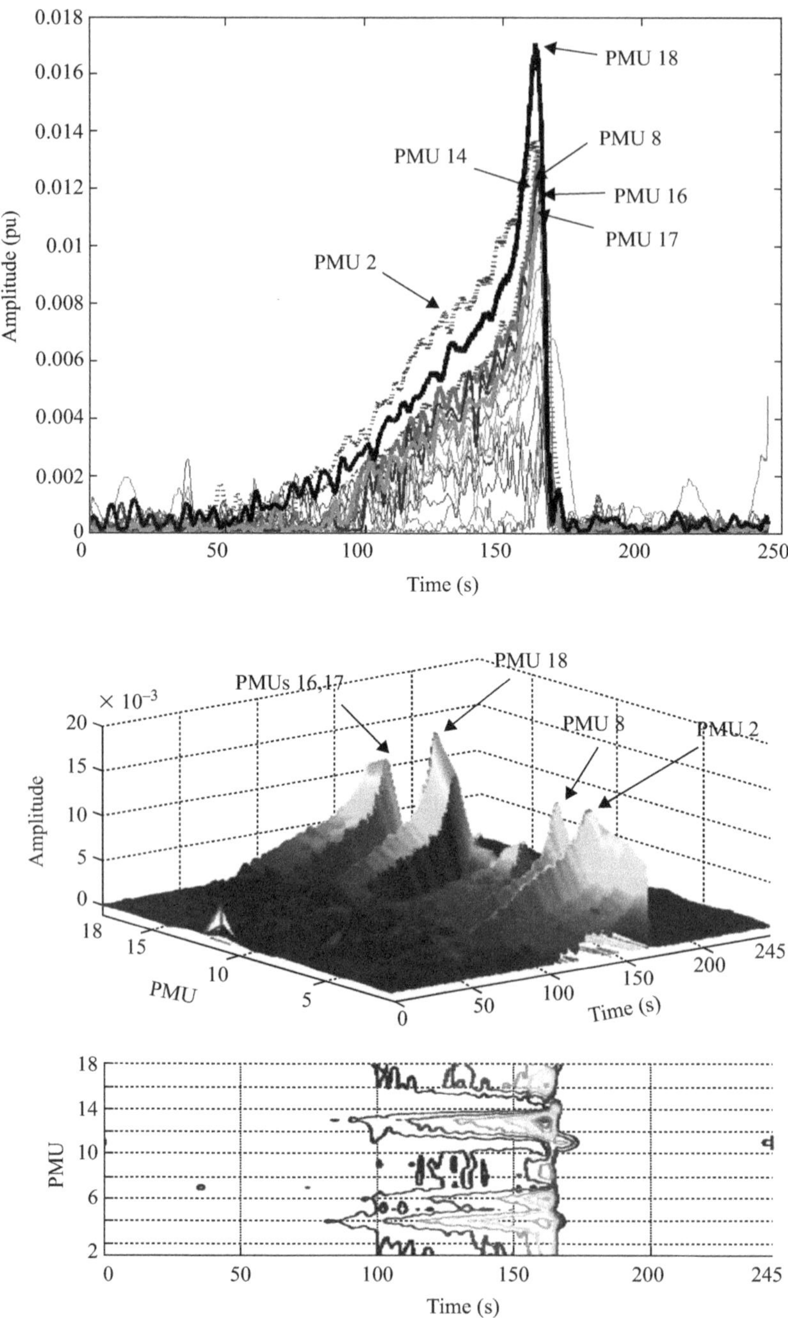

Figure 8.10 Spatio temporal pattern associated with the 1.0 Hz mode. (a) Time-frequency-location representation, and (b) Projection onto the time-PMU plane

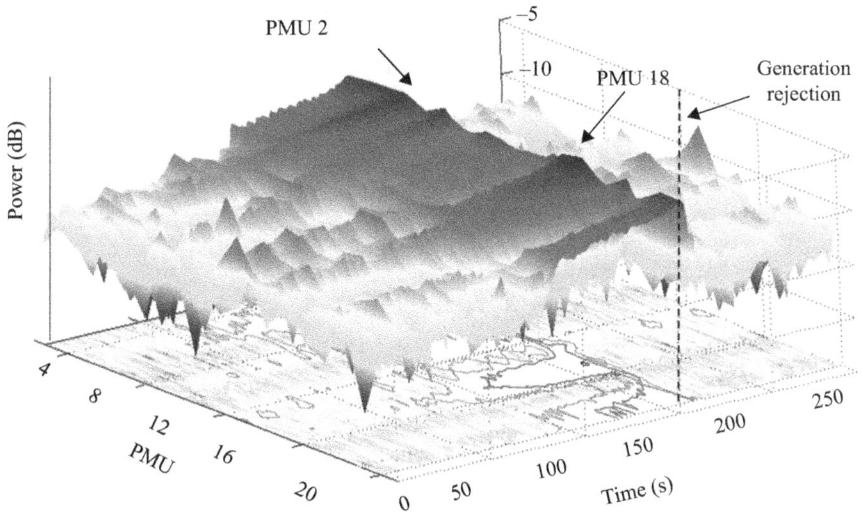

Figure 8.11 Spatio temporal pattern associated with the 1.0 Hz mode using Sensor-level fusion in (8.4)

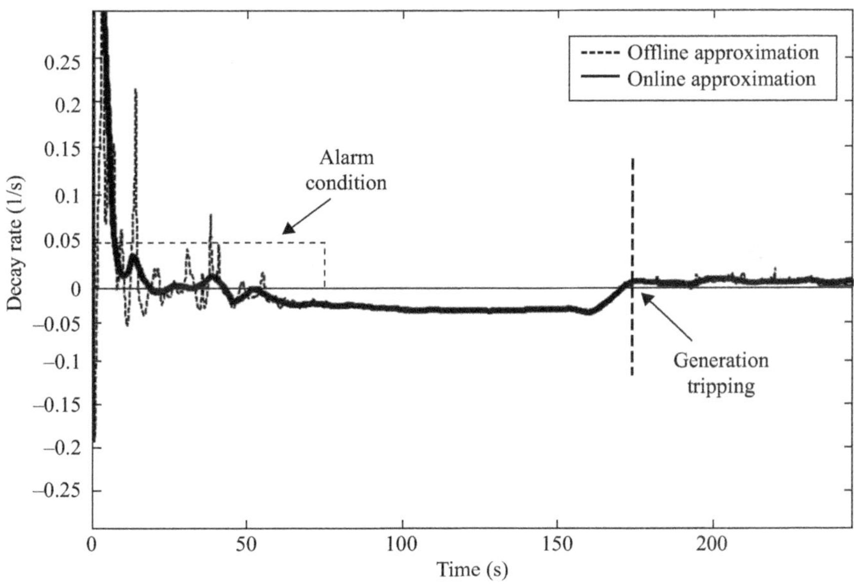

Figure 8.12 Decay rate of IMF 2

Two analytical approaches were assessed and compared:

1. The conventional offline formulation in Chapter 4
2. A recursive implementation of the method in Chapter 7. For this analysis a 6 s window with no overlapping was used (refer to Figure 7.14)

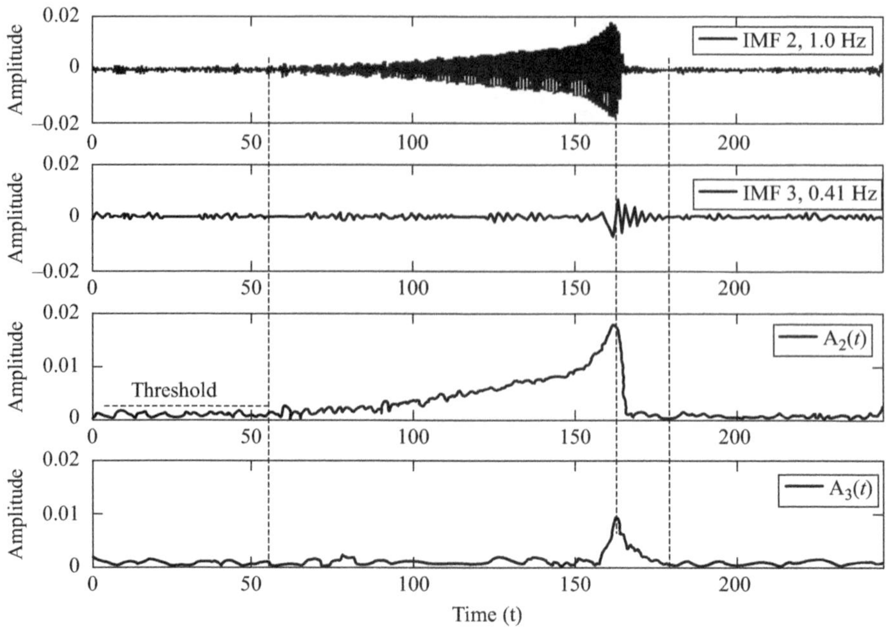

Figure 8.13 Instantaneous amplitude and IMFs

In the latter case, HHT analysis was applied to sliding windows of the form

$$\mathbf{X}_j^k(t_{w_k}) = \begin{bmatrix} c_{1j}^k(t_1) & c_{1j}^k(t_2) & \cdots & c_{1j}^k(t_N) \\ c_{2j}^k(t_1) & c_{2j}^k(t_2) & \cdots & c_{2j}^k(t_N) \\ \vdots & \vdots & \ddots & \vdots \\ c_{m_kj}^k(t_1) & c_{m_kj}^k(t_2) & \cdots & c_{m_kj}^k(t_N) \end{bmatrix}, \quad j = 1, ..., p_k$$

where the index k denotes the current window, and $p_k = 1$ for the analysis of the 1.0 Hz mode.

Damping estimates are obtained from

$$\sigma_j(t) = \text{Re}\left[\dot{z}_j(t)/z_j(t)\right]$$

where $z(t) = c(t) + jc_H(t)$.

An approximate estimate can also be obtained by noting that the time-dependent amplitudes $A_j(t)$ are typically of the general amplitude-modulated (AM) form

$$A_j(t) = e^{-\sigma_j(t)} \cos(A_o + A_{am} \cos \omega_{am} t) \tag{8.5}$$

where A_o and A_{am} represent the coefficients of the frequency-modulated (FM) signal

Taking the natural logarithm of (8.5) results in

$$\sigma t = \ln(A_o) + \ln(A_{am} \cos(\omega_{am} t) - \ln a(t)$$

where, in the more general and interesting case, $A_{am} \cos(\omega_{am} t)$ is an unknown quantity. It follows that

$$\sigma = \frac{d}{dt}[\ln(A_o + A_{am} \cos \omega_{am} t)] - \frac{d}{dt} \ln a(t) \tag{8.6}$$

Equation (8.6) reduces to the usual definition $\sigma t = \ln(A_o) - \ln a(t)$ for the case in which amplitude modulation is absent ($A_{am} = 0$).

Damping estimates in Figure 8.12 show that the online method accurately identifies the exact time at which the system becomes unstable at about 55 s. Simulation results in Figure 8.13 show that both energy (entropy) and damping can be used to trigger control actions. Energy can also be used to analyze the onset of specific system behavior associated with specific modal behavior. Results indicate the exact time in which the various system modes are excited.

Of particular interest for the analysis, results show that the system is made stable when the critical plant trips out at about 175 s (refer to Figures 8.3 through 8.5).

Careful inspection of Figure 8.12 shows that near real-time approximations result in a smother and more accurate representation of system damping. Drawing on these ideas, an alarm and triggering system were designed based on damping information for use in emergency control applications. Discussion is deferred to section 8.7.

In practice, detection of deterioration conditions may be limited by high noise levels in the measurements, especially under ambient conditions. To enhance the accuracy of modal estimates, techniques to subtract the higher frequency components in the signal based on recursive application of the EMD/wavelet techniques in section 7.4 can be used.

In the following, the local HHT estimates in section 8.3.3 are compared to wide-area measurements with regard to their ability to capture the full system dynamics.

8.3.4 Multitemporal, multiscale analysis of measured data

In this analysis, system dynamic behavior associated with data from regional PDCs was analyzed using partial least squares and PCA. For ease of comparison, the northern system exhibiting the largest frequency deviations is taken as a reference for modal analysis.

Based on the actual PDCs deployed in the system (refer to Table 8.2), the following observation matrices were selected for analysis:

1. Northern system:

$$\mathbf{X}_j^1(t) = \begin{bmatrix} c_{1j}^1(t_1) & c_{1j}^1(t_2) & \cdots & c_{1j}^1(t_N) \\ c_{2j}^1(t_1) & c_{2j}^1(t_2) & \cdots & c_{2j}^1(t_N) \\ \vdots & \vdots & \ddots & \vdots \\ c_{m_1j}^1(t_1) & c_{m_1j}^1(t_2) & \cdots & c_{m_1j}^1(t_N) \end{bmatrix}$$

2. Northeastern system:

$$\mathbf{X}_j^2(t) = \begin{bmatrix} c_{1j}^2(t_1) & c_{1j}^2(t_2) & \cdots & c_{1j}^2(t_N) \\ c_{2j}^2(t_1) & c_{2j}^2(t_2) & \cdots & c_{2j}^2(t_N) \\ \vdots & \vdots & \ddots & \vdots \\ c_{m_2j}^2(t_1) & c_{m_2j}^2(t_2) & \cdots & c_{m_2j}^2(t_N) \end{bmatrix}$$

3. Southeastern system:

$$\mathbf{X}_j^3(t) = \begin{bmatrix} c_{1j}^3(t_1) & c_{1j}^3(t_2) & \cdots & c_{1j}^3(t_N) \\ c_{2j}^3(t_1) & c_{2j}^3(t_2) & \cdots & c_{2j}^3(t_N) \\ \vdots & \vdots & \ddots & \vdots \\ c_{m_3j}^3(t_1) & c_{m_3j}^3(t_2) & \cdots & c_{m_3j}^3(t_N) \end{bmatrix}$$

with $m_1 = 5$, $m_2 = 12$, $m_3 = 1$, and subscript $j = 1$ refers to the 1.0 Hz mode. For PDCs 1 and 2, values above represent multisensor data.

Preserving the number of measurements, the overall observation matrix, $\mathbf{X}_f$, can now be rewritten in the form

$$\mathbf{X}_f = \begin{bmatrix} \mathbf{X}_j^1 \\ \hline \mathbf{X}_j^2 \\ \hline \mathbf{X}_j^3 \end{bmatrix} \tag{8.7}$$

This case corresponds to the hybrid multiblock PCA (POD) analysis in Figure 5.9. Table 8.4 shows the computational effort for partial least-squares analysis. As shown, partial least-squares analysis of two-block systems drastically reduces CPU time while accurately extracting the relevant behavior of interest.

Other wide-area measurement system (WAMS) architectures were tested but did not perform well in comparison and are not discussed here.

Figure 8.14 shows the scores for the combined analysis of northern and northeastern systems. As noted in previous sections, score 1 approximates the

Table 8.4 CPU time: Full system record

Technique	CPU time (s)	Observations
Linear PCA/POD	0.2000	Whole time interval, 18 signals analyzed simultaneously
Multiblock PCA (northern system, northeastern system)	0.0027	Whole time interval, 17 signals analyzed simultaneously
Multiblock PCA (northeastern system, southeastern system)	0.0022	Whole time interval 13 signals analyzed simultaneously

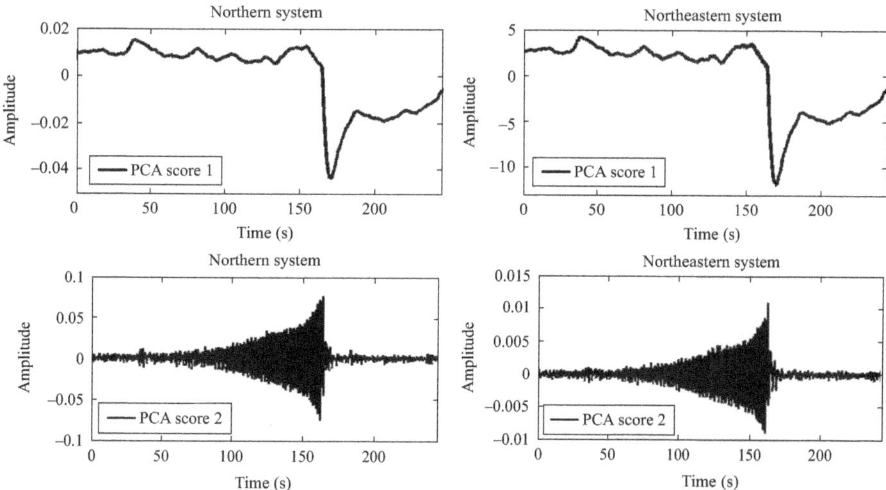

Figure 8.14 PLS analysis of PDC data from the northern and northeastern system

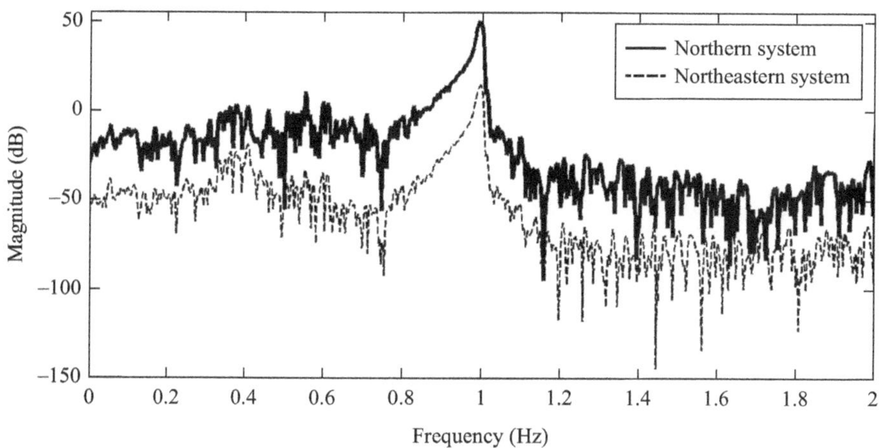

Figure 8.15 Spectra of PDC data from the northern and northeastern systems

system trend, whereas score 2 approximates the dominant 1.0 Hz component in agreement with the results from other techniques.

Figure 8.15 depicts the spectra of PCA score 2 in Figure 8.14, while Figure 8.16 shows the corresponding loadings from PLS analysis. Results are found to be consistent with conventional PCA analysis in Figure 8.14.

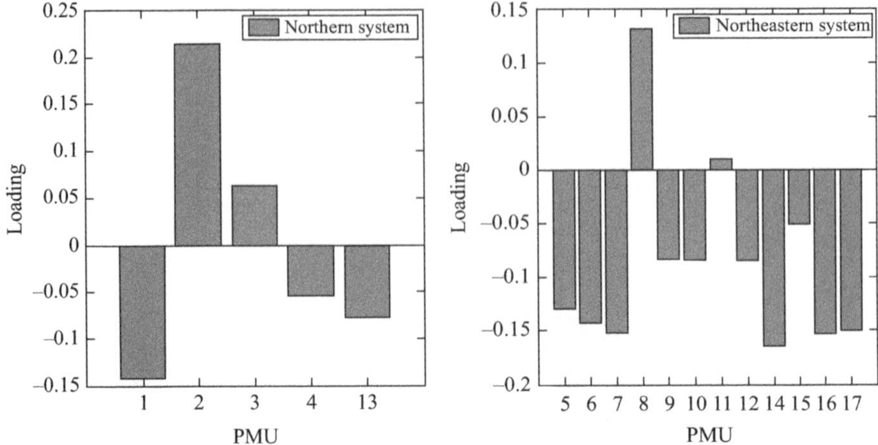

Figure 8.16 Loadings from PLS analysis

Table 8.5 Regression coefficients for the two-block northern–northeastern system analysis

NE/N	PMU 1	PMU 2	PMU 3	PMU 4	PMU 5	PMU 6
PMU 1	−0.0114	0.2435	0.1978	0.4563	0.3053	−0.2029
PMU 2	−0.0020	−0.2096	−0.6807	1.1069	0.5798	0.2035
PMU 3	−0.0042	0.1704	0.0213	0.6380	0.1405	0.0297
PMU 4	−0.0059	0.1671	0.0127	0.6493	0.1374	0.0333
PMU 5	0.0013	0.0029	−0.2893	0.8479	0.3278	0.1105
PMU 6	−0.0069	0.1691	0.0174	0.6438	0.1362	0.0334
PMU 7	−0.0314	0.2450	0.3134	0.1252	0.5724	−0.2557
PMU 8	0.0106	0.4750	0.6967	−0.0935	−0.0815	0.0030
PMU 9	0.0005	0.0116	−0.1720	0.5628	0.6671	−0.0695
PMU 10	−0.0230	0.2618	0.3951	−0.0430	0.6761	−0.2897
PMU 11	−0.0013	0.1655	0.0545	0.5127	0.2051	0.0619
PMU 12	−0.0060	0.1602	−0.0911	0.9388	0.0038	−0.0116

From (5.12), the relation between the PMU measurements in the northern and northeastern systems is given by

$$\mathbf{U} = \mathbf{B}\mathbf{T} + \mathbf{R}_u \tag{8.8}$$

where **B** is the *m*-by-*P* matrix of scores from the partial least-squares decomposition.

Table 8.5 gives the coefficients **B** for the analysis above. A clear pattern among the coefficients can be seen by comparing the entries of Table 8.6.

Plots of estimated residuals are shown in Figure 8.17 for the northern system.

Figure 8.18 shows the corresponding PLS analysis for the northeastern and southeastern systems.

Table 8.6 Beta: Northeastern–southeastern

Northeastern/Southeastern	PMU 1
PMU 5	−0.0057
PMU 6	0.3529
PMU 7	0.5671
PMU 8	−0.6617
PMU 9	1.1666
PMU 10	−0.6786
PMU 11	−0.5597
PMU 12	0.1703
PMU 14	−0.5551
PMU 15	0.4460
PMU 16	1.0728
PMU 17	−0.6133

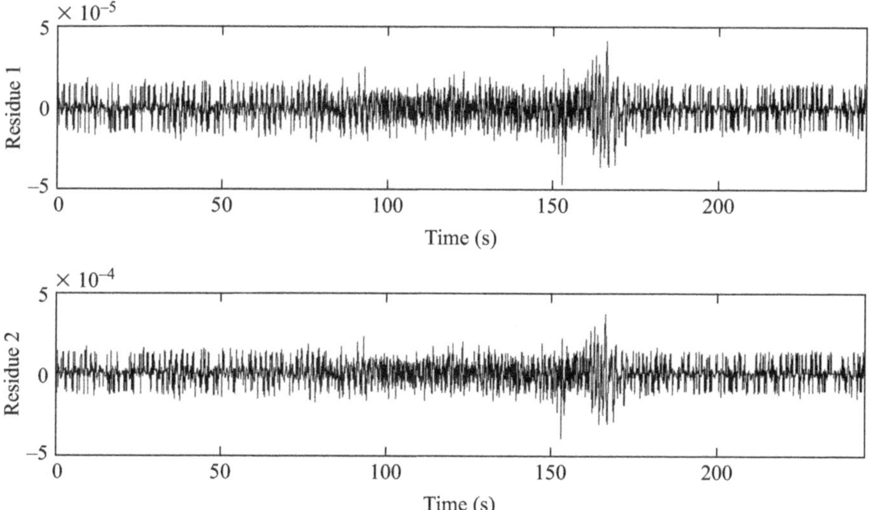

Figure 8.17 Behavior of PLS residuals: northern system

8.3.4.1 Multimodal data

Recorded dynamic data for the above event include voltage, frequency, power, and phase angle measurements. In the numerical analysis described in section 8.3.5 72 signals were used for modal characterization. For purposes of analysis, data was normalized and detrended (PCA/POD, Prony).

8.3.5 Performance evaluation

In this section the performance of the above techniques is compared in terms of accuracy and CPU effort. Based on the theoretical modes in previous chapters,

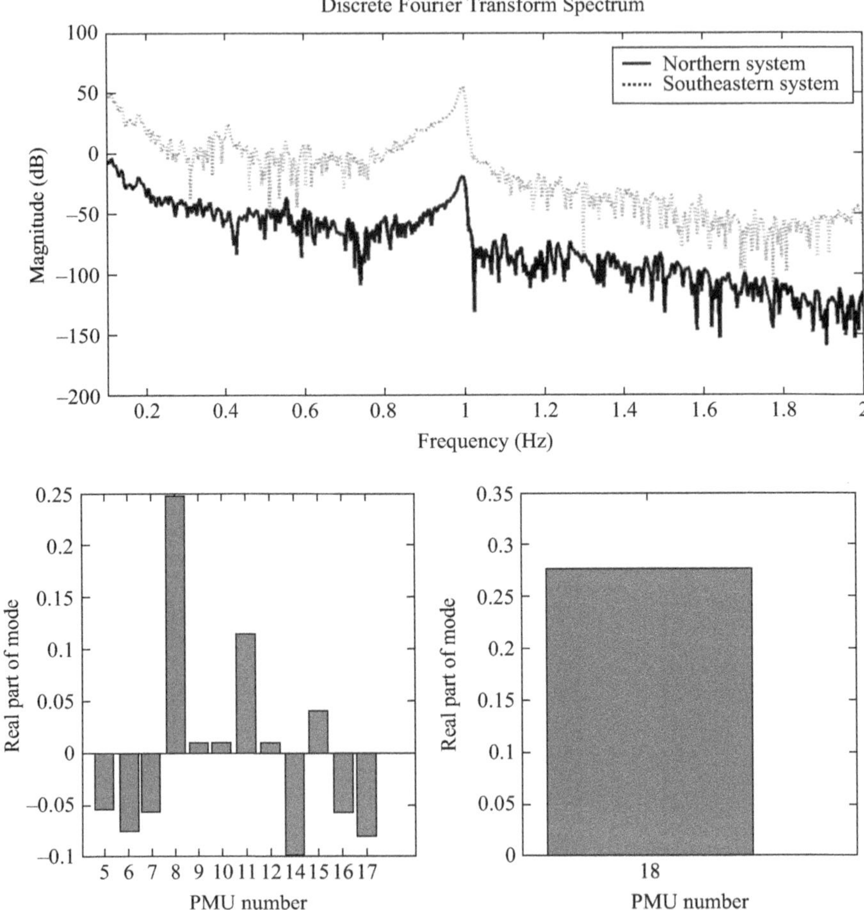

Figure 8.18 PLS analysis of PDC data from the northeastern and southeastern systems

Table 8.7 CPU time for modal characterization

Technique	CPU time	Observations
POD	0.2000	Computation of modal components
BSS	0.0000	Computation of modal components
Diffusion analysis	0.0375	Computation of k-diffusion maps + time coefficients

MATLAB codes for the various methods were developed and tested. These results are only illustrative, since no efforts were made to optimize the codes.

Tables 8.7 and 8.8 show the CPU time needed to characterize modal behavior for the frequency signals in Figure 8.2. These results should be compared with the application of other global monitoring techniques in [14].

Table 8.8 CPU time for various modal estimation methods

Technique	Prony	Koopman	POD	Diffusion[b]
0–120 s	[a]	65.35 s	0.096 s	0.0375
120–160 s	3.73 s	1.395 s	0.099 s	0.0323

[a]No physically meaningful solution was obtained.
[b]Extraction of diffusion coordinates.

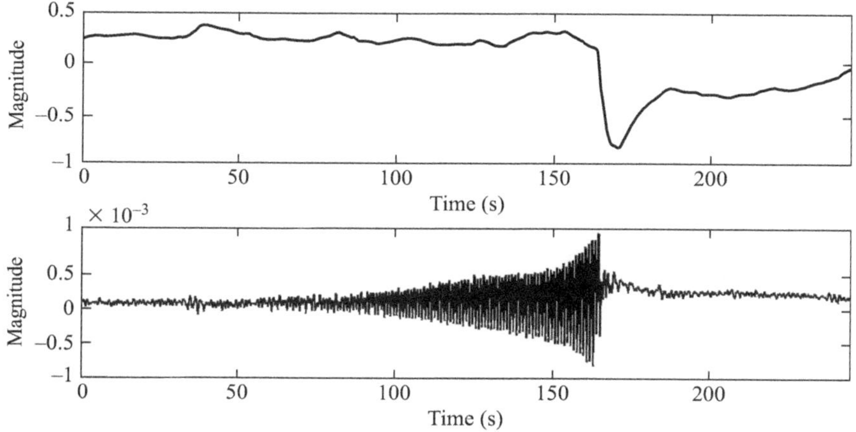

Figure 8.19 Temporal coefficients $a_o(t)$ and $a_o(t)$

As a further test, the methods were used to simultaneously analyze the 72 records of dynamic data. Representative simulation results using conventional diffusion maps and PCA for 18 000 samples are as follows:

- Diffusion map analysis, 1.53 s
- PCA, 2.23 s

These results are compared favorably with Prony and Koopman analyses in Table 8.8 for extracting dynamic patterns (mode shape and clustering information).

8.4 Pattern recognition analysis

Further information about the nature of transient behavior is obtained from the application of high-dimensional pattern recognition techniques. In this analysis, the raw measurements (8.2) were analyzed using the diffusion analysis.

8.4.1 Diffusion map analysis

In the first step, the distance matrix was computed from the 18 dynamic trajectories. To understand the nature of the underlying system behavior Figure 8.19

shows a plot of the time-dependent coefficients $a_o(t)$ and $a_1(t)$ extracted using the relations

$$\begin{aligned} a_o(t) &= \mathbf{X}\boldsymbol{\Psi}_1 \\ a_1(t) &= \mathbf{X}\boldsymbol{\Psi}_2 \end{aligned} \tag{8.9}$$

These results are typical. As shown, the method is able to separate the slow motion (system trend) from the oscillatory behavior associated with the dominant modes.

Figure 8.20a shows the eigenvalues λ_k in decreasing order of variance. As in previous analyses, the eigenvalues are ordered in such a way that the first eigenvalue corresponds to the mean motion, the second to the main oscillatory behavior, etc.

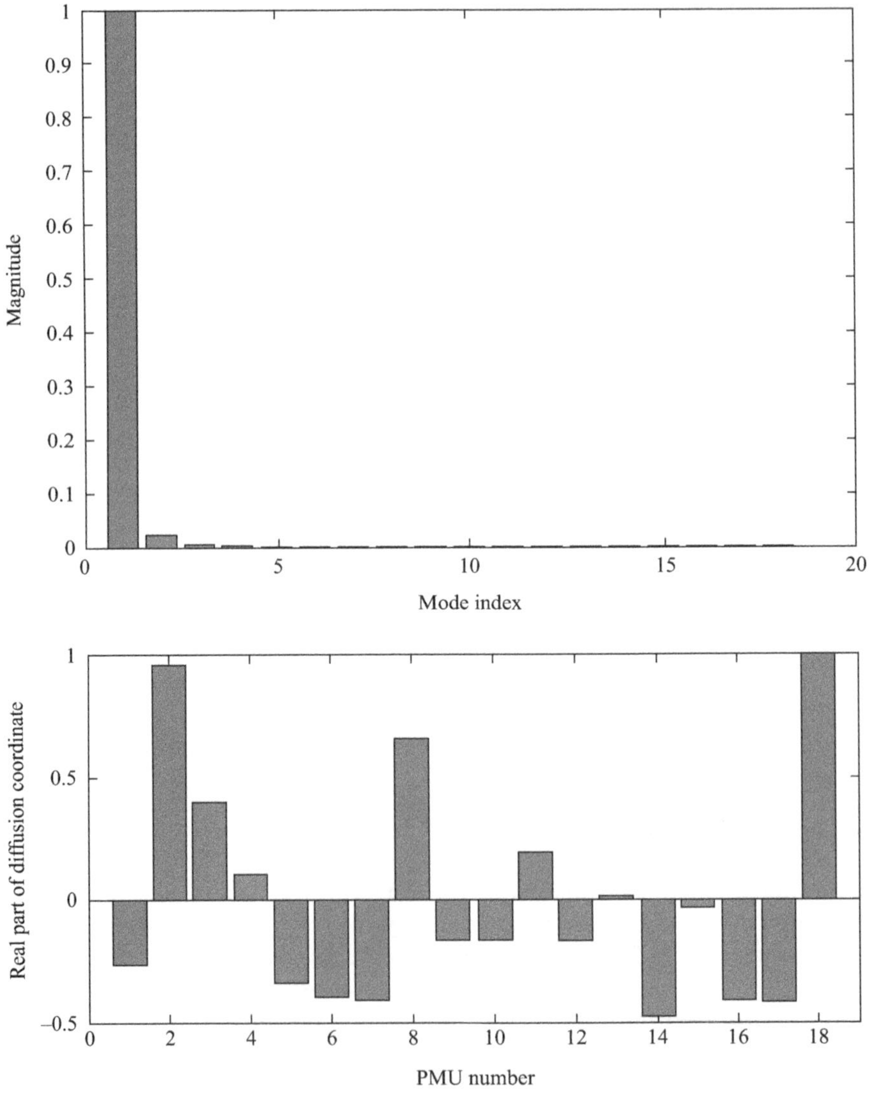

Figure 8.20 Diffusion map: (a) eigenvalue spectra; (b) diffusion coordinate, f_2

Eigenvalues whose magnitude is smaller than a given threshold do not explain much spatial variation and can be excluded of the analysis.

Two modes are seen to capture nearly 99% of the total energy. As pointed out above, the first singular value represents the average system behavior, while the second temporal mode captures the oscillatory behavior.

Further insight into the nature of system behavior can be gleaned from the analysis of the diffusion coordinates in Figure 8.20b. Comparison of this plot, with the corresponding spatial patterns in Figure 8.7, shows that the diffusion vectors accurately capture the spatial behavior of the signals.

The diffusion map identifies PMUs 18, 2, and 8 as those that have the largest amplitude. These records correspond to the bus frequency deviations exhibiting the largest frequency deviations in Figure 8.3.

To further verify the accuracy of the nonlinear dimensionality reduction technique, the 18 frequency signals were denoised and detrended simultaneously using the approach in section 7.3.1 using the following two-stage approach.

1. In the first stage, the signals were denoised using the multiscale wavelet denoising technique.
2. In the second stage, the signals were demeaned.

While this is not necessary in the application of the method, the analysis allows correlation of the obtained results with observed data. Careful analysis of Figure 8.21

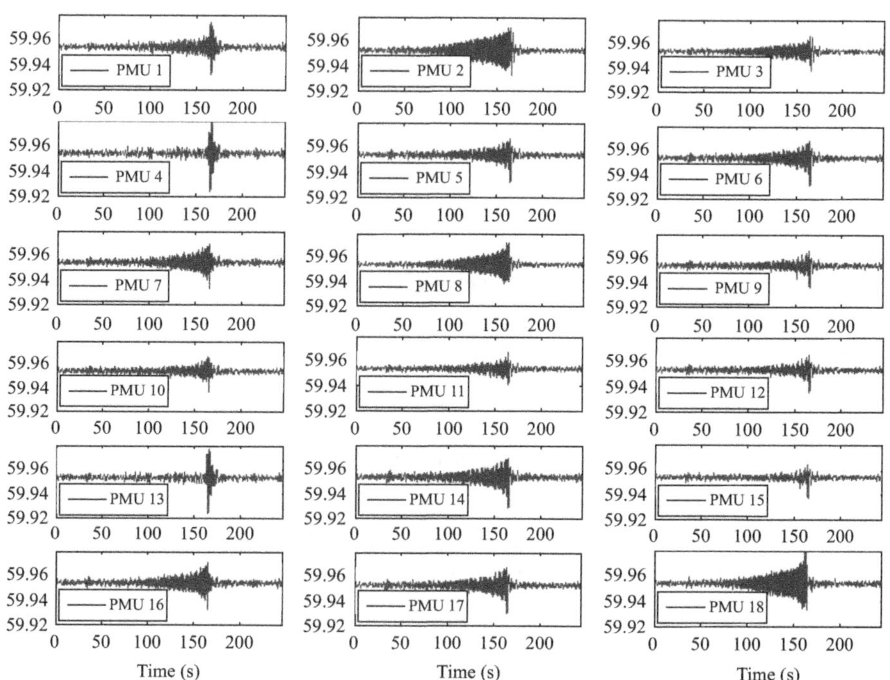

Figure 8.21 Detrended and denoised PMU measurements: Plots describe denoised and detrended signals associated with the 1.0 Hz mode

allows confirmation that signals 18, 2, 7, 17, and 16 show the largest deviation from nominal behavior.

To further verify the appropriateness of the analysis, Prony analysis was performed on the oscillatory component, $a_2(t)$. Three time intervals are considered in the analysis: (a) 0–120 s, (b) 120–160 s, and (c) 180–220 s. Estimates of the modal damping and frequency obtained using Prony analysis are shown in Table 8.9.

These results illustrate several advantages of the application of nonlinear modal reduction techniques. Values in Table 8.9 show that the diffusion-based approach accurately estimates frequency and damping with low computational effort. In addition, the extracted diffusion coordinates provide additional information on energy exchange.

8.4.1.1 Identification of coherent groups

As discussed in Chapter 3, clusters of coherent signals can be identified from the analysis of diffusion maps. Table 8.10 presents the frequency grouping obtained using the *k*-means clustering algorithm. In these studies, six clusters were selected for analysis. The results match very well observed behavior.

8.4.2 Comparison with other approaches

To independently verify the accuracy of the modal approximations, other popular nonlinear model reduction techniques were used to extract mode shapes from raw measurements.

Figure 8.22 shows the extracted frequency-based mode shapes from measured data using two distinct analytical approaches:

1. The Isomap method
2. Laplacian eigenmap analysis

Table 8.9 Prony analysis results on $a_2(t)$

Time interval	*Frequency* (Hz)	*Damping* ($\xi/2\pi$)
0–120 s	1.008	−0.018
120–160 s	0.995	−0.023
180–220 s	0.998	−0.031

Table 8.10 k-means clusters

Cluster	PMU	Area
1	9, 10, 12, 15, 16, 7	Northern
2	3, 8, 11	Northeastern
3	5, 6, 17, 14	Northeastern
4	2	Northern
5	18	Southeastern
6	1, 4, 13	Northern

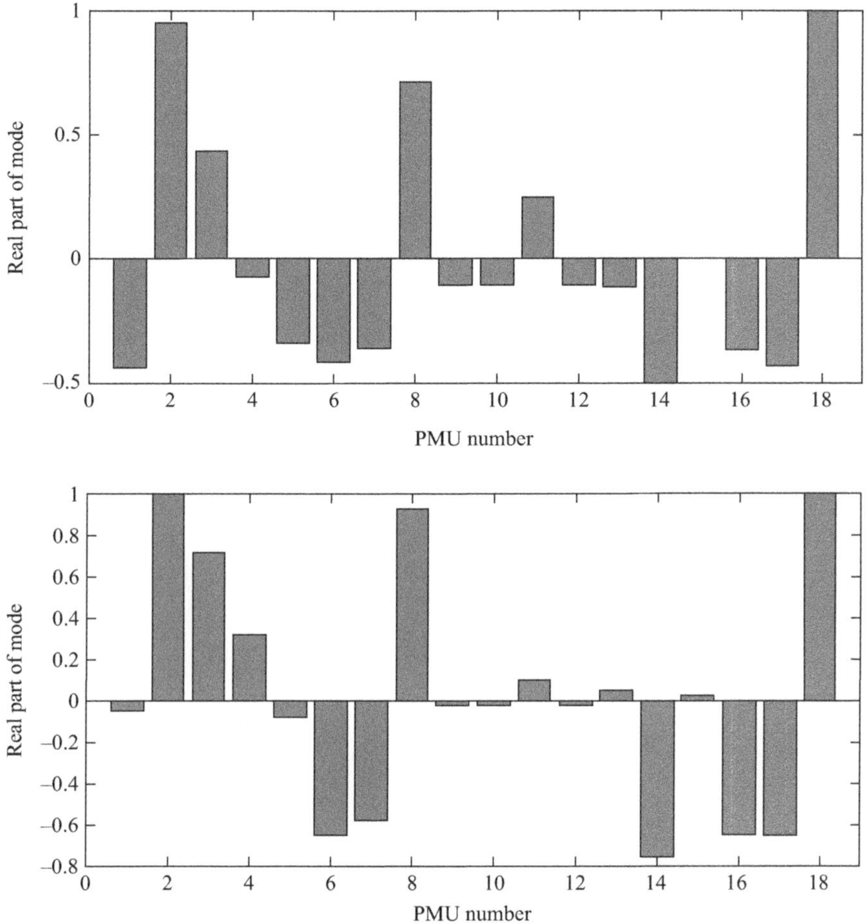

Figure 8.22 Extracted mode shapes from measured data: (a) Isomap; (b) Laplacian eigenmap

In the first case, the error between the pair-wise distances in the low-dimensional ($\mathbf{Y} = \{y_1, \ldots, y_d\}$) and high-dimensional ($\mathbf{X} = \{x_1, \ldots, x_n\}$) representations of the data were minimized using the objective function

$$J_{Iso} = \sum \left(\|x_i - x_{ij}\| - \|y_i - y_{ij}\| \right)^2$$

A pseudo-Newton method was used in the calculations.

In the second case, the cost function that is minimized is given by

$$J_{Lap} = \sum \left(\|y_i - y_j\| \right)^2 w_{ij}$$

where the w_{ij} are weight coefficients.

Again, results compare well with modal information in sections 8.3 and 8.4.

8.5 POD/BSS analysis

In this analysis, the POD/PCA-based monitoring technique was applied to a combination of fields representing PMU measurements. The observational data from sensors was arranged into an n-by-N modal response data, with each row representing time-series data of an individual sensor, that is

$$\mathbf{x}(t_j) = \sum_{i=1}^{p} a_i(t)\,\varphi_i(x), \quad j = 1, ..., N \tag{8.10}$$

Figure 8.23a displays the significant POMs for the analyzed data, whereas Figure 8.23b shows the EOFs obtained by removing the mean value. Modes 1 and 2 explain, respectively, 92% and 8% of the total energy.

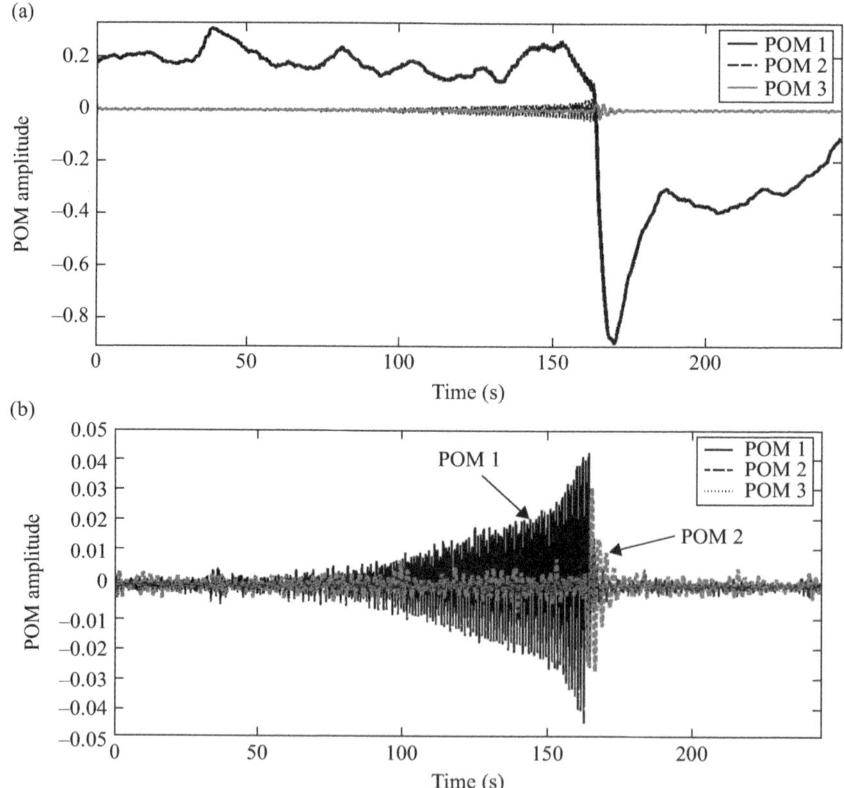

Figure 8.23 Time evolution of dominant POMs: (a) raw measurements; (b) detrended signals

As suggested in Figure 8.23a, the first EOF (POM 1) represents the average value of the physical variables associated with measurements, i.e.

$$a_1(t) = POM_1(t) = f_{ave}(t) = \frac{1}{N}\sum_{j=1}^{N} f_j(t) \tag{8.11}$$

in agreement with the average frequency representation.

POM 2 in Figure 8.23a, in turn, is seen to capture the prominent system variability and is used for real-time monitoring of system behavior.

Also of interest, Figure 8.24 compares the time evolution of IMF 1 to the average frequency deviation as given by (8.11). As shown, the results are undistinguishable showing the accuracy of the model.

Finally, comparison of the Hilbert spectra of EOF 1 in Figure 8.25 with the spectra of the frequency recording of PMU 13 shows that the low-order representation accurately captures the main features of interest in the process.

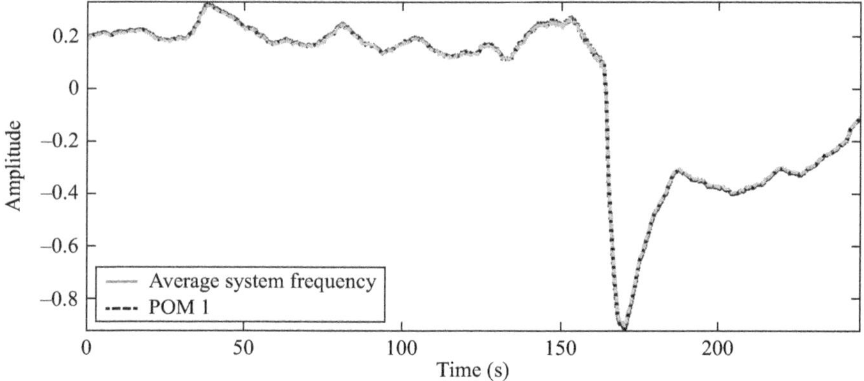

Figure 8.24 Comparison of POM 1 with the average system frequency

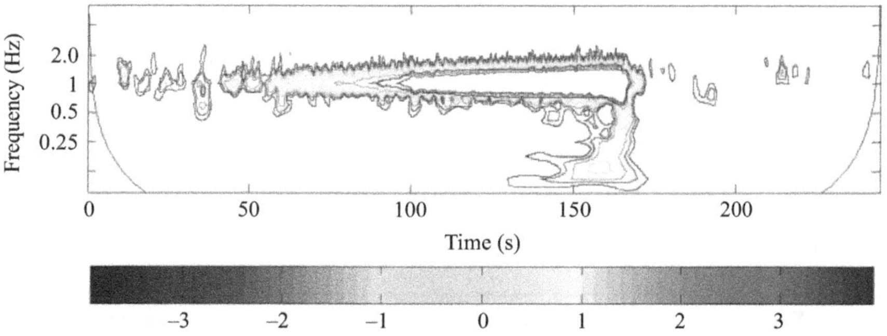

Figure 8.25 Hilbert spectrum of POM 1

8.6 Validation of power system model

The sequence of generator and line tripping in Table 8.1 was simulated using a detailed database of the system. These simulations were used to validate the system model against the collected measurements.

A detailed transient stability model replicating operating conditions for the above event was to this end developed and tested: the sequence of events was simulated using a detailed dynamic database of the system.

The model includes the following:

- 5 000 buses with 550-generator 2 245 loads and several large static VAR compensators
- Detailed excitation system models and turbine governors
- Eight areas of the system modified to represent the pre-event power flow conditions using available supervisory control and data acquisition (SCADA) and PMU data

8.6.1 Small signal performance

Detailed small-signal analyses were conducted to further verify the accuracy of the system model. Table 8.11 displays the main characteristics of the slowest inter-area modes in the system. Among these modes, the 0.42 Hz and 0.62 Hz modes involving the interaction of machines in the north and south systems are of particular interest here.

Figure 8.26 also shows extracted speed-based mode shapes for the 0.42 Hz and 1.0 Hz modes in the system.

8.6.2 Large system performance

Figure 8.27 shows a comparison of the observed (measured) and simulated frequency deviations for time interval 3 in Figure 8.1. Figure 8.12a shows large-scale time domain simulation for the plant outage event in Table 8.1. Figure 8.17c shows the corresponding Fourier spectra. The plots show that the response of the simulated system model agrees with the disturbance system recordings shown in Figure 8.27b.

As shown in this plot, the frequency drops at about 59.78 Hz in close agreement with measured data in Figure 8.3. The fast Fourier transform (FFT) spectra, however, show dominant inter-area modes at about 0.40 Hz, 0.69 Hz, and 0.84 Hz associated with dominant inter-area modes in the system expected from measured data.

Both the dominant bus frequencies and the swing patterns are accurately determined.

Table 8.11 Slowest inter-area modes in the system

Mode	Eigenvalue	Frequency (Hz)	Damping (%)	Dynamic pattern
1	$-0.1414 \pm j2.655$	0.4226	5.32	South systems vs. north systems
2	$0.3308 \pm j6.464$	1.0288	−5.16	Generators 325, 326, and 327 vs. the rest of the system

(a)

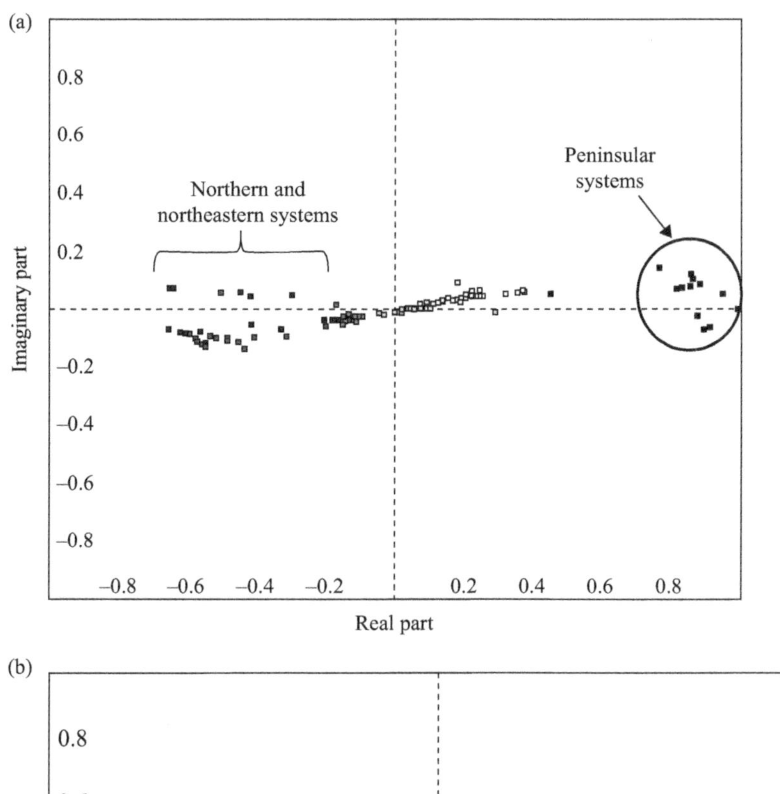

(b)

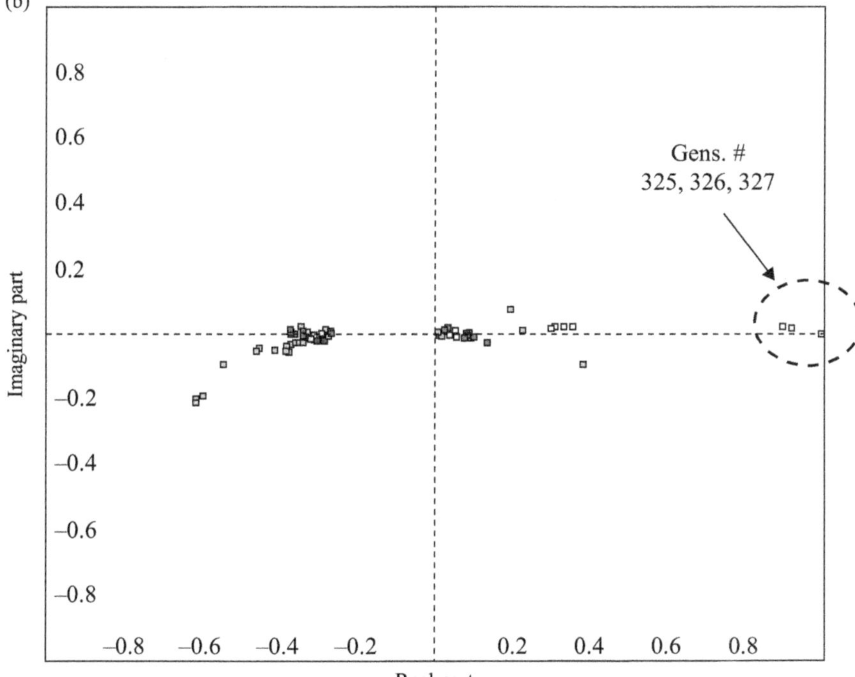

Figure 8.26 Speed-based mode shape of relevant system modes: (a) 0.42 Hz mode; (b) 1.0 Hz mode

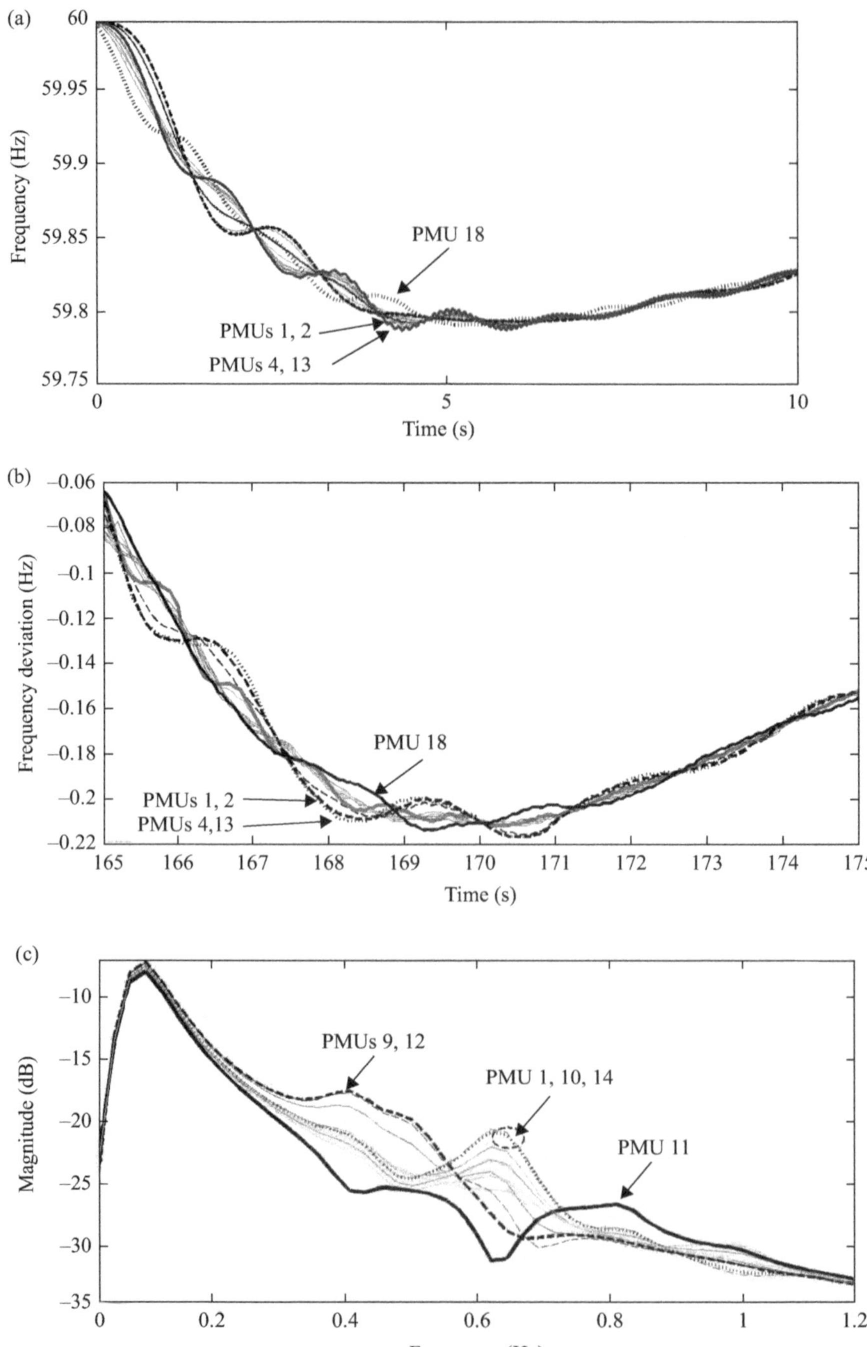

Figure 8.27 Simulated bus frequencies for time interval 2 in Figure 8.1. Base case. (a) Frequency traces, time window 3. (b) Measured data; frequency deviations are shown relative to the nominal 60 Hz frequency. (c) Spectra of simulated data

8.7 Evaluation of control performance

Drawing on the above analysis, a coordinated, damping-based wide-area control scheme was designed and tested using the validated large-scale power system model. Emphasis was placed on the ability of the technique to enhance the post-disturbance response of the system (time window 3).

The proposed wide-area monitoring and control (WAMC) system integrates sensor networks and monitoring structures to monitor and control system health and may be divided into three levels:

1. *Multivariate EMD/EOF-based modal estimation:* In this stage, a multivariate POD/EMD technique is used to identify inter-regional oscillatory swings that pose the highest risk, based upon synchronized phasor measurements.

   For monitoring purposes, the HHT is applied online to a moving window of adjustable length and the dominant scales are determined using the approach discussed in Chapter 7. An EOF, data-driven statistical approach is used to extract global features from the modal response matrix.

   This estimator allows to determine the location and extent of power system degradation by tracking modal properties in near-real time. The output of this module represents an approximation of modal properties at selected system locations.

   Once the most energetic modes are identified, the most critical buses (and their associated geographical locations) are identified using modal amplitudes and mode shapes. Note that because the complex information contains information about the phase of the oscillation, the coherent bus groups are determined.
2. *Adaptive, damping-based triggering:* Based on modal properties in step 1, a triggering algorithm is used to set a logic flag when instantaneous damping of the most energetic modes decreases below a certain value, that is, 3% for a given number of cycles.

   Experience shows that the combined application of damping and entropy information sharpens the ability of the technique to identify and isolate the slowest dynamics associated with critical inter-regional oscillations.
3. *Wide-area remedial action schemes:* To control system health, various remedial action schemes have been implemented. This scheme issues triggering commands to damping-based control action schemes at selected system locations based on the information in stages 1 and 2.

A schematic illustration of the proposed methodology is presented in Figure 8.28. Damping control actions currently used in the MIS include load shedding schemes, capacitor switching, and damping-based control strategies designed to activate the power damping control mode (modulation control) in Flexible AC Transmission Systems (FACTS) operating in voltage regulation/impedance control mode. Alternatives being investigated include static VAR compensators (SVCs) and controlled series capacitors.

The general control structure being investigated is shown in Figure 8.28. As shown in this plot, the damping control action, adapted for controlling one or more

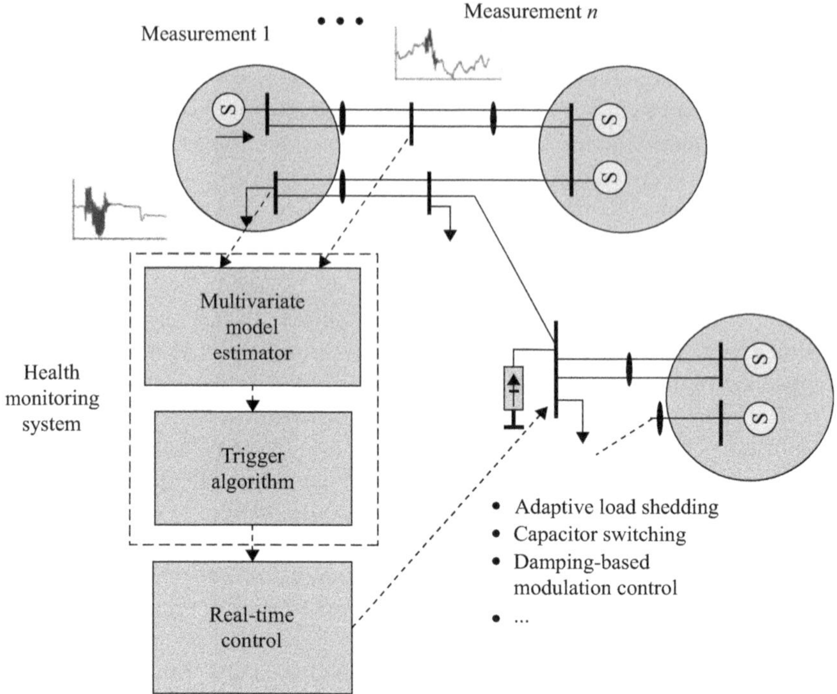

Figure 8.28 Conceptual view of the proposed algorithm

critical inter-area modes, is initiated in response to a system threat manifested by a deterioration of modal damping.

Two control strategies were considered in the exploratory studies:

1. A fast wide-area load shedding scheme. In this scheme, load shedding is initiated when the instantaneous damping of critical frequency measurements drops below 3% for various cycles.
2. A damping-based modulation control using the control structure shown in Figure 8.29.

Figure 8.28 shows a schematic diagram of the proposed framework. The framework includes three main steps: global monitoring of system behavior, an entropy-based triggering algorithm, and a near real-time wide-area control system. The output of this scheme is a triggering signal to a nearby SVC or a load shedding action. Each alternative was simulated using the same base case.

The health monitoring system constantly monitors the system operating status and issues the appropriate triggering actions for tripping or initiating modulation controls.

Figure 8.30 shows simulation results for tripping load at critical system locations close to PMU 12 and SVC modulation control at PMU 6 following generation tripping (refer to Figure 8.28).

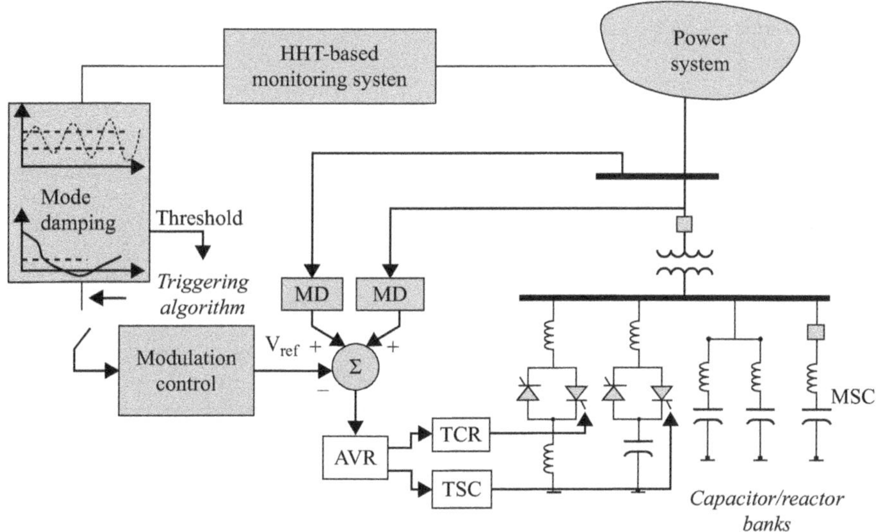

Figure 8.29 SVC control block representation

For simulation, load shedding was represented on 70% of the load and the initial frequency is assumed to be 60 Hz. A general transfer function of the form

$$G_{\text{mod}}(s) = K_{psdc}\frac{sT_w}{1+sT_w}\frac{(1+sT_1)(1+sT_2)}{(1+sT_3)(1+sT_4)}\frac{T_m}{1+sT_m}$$

was used in the simulations [15].

Comparison with the base case in Figure 8.27a shows that the corrective control actions effectively reduce the minimum post-contingency frequency and the settling frequency before AGC action.

To further verify the ability of the proposed control strategy, modal damping for the post-contingency operating condition was computed using Prony analysis. Using this technique, the post-disturbance measurements are expressed in the form

$$x(t) = \sum_{j=1}^{q} A_i e^{\sigma_j t}\cos(\omega_j t + \phi_j) \tag{8.12}$$

where A_j, σ_j, ω_j and φ_j are the amplitude, damping, frequency, and phase of the *j*th modal component.

Tables 8.12 and 8.13 summarize the results of the various control strategies. As shown, control cases demonstrate a significant improvement over the base case damping condition.

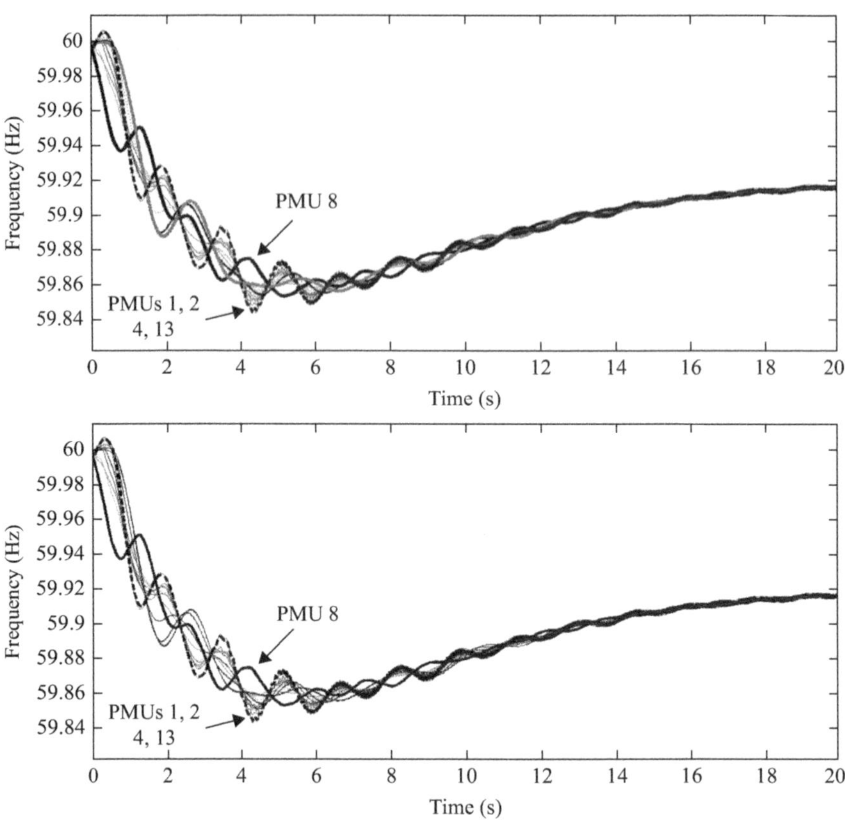

Figure 8.30 Simulated bus frequencies for time window 2 in Figure 8.3. Adaptive wide-area remedial schemes activated. (a) Load shedding near PMU 13. (b) Combined load shedding near PMU 12 and modulation control at the SVC at PMU 6 bus.

Table 8.12 Prony analysis fit: frequency measurement at PMU 12 (post-contingency condition)

Control alternative	Amplitude	Frequency (Hz)	Damping (%)
Base case	0.0062	0.414	6.77
	0.0217	0.653	5.01
	0.0095	0.846	5.69
Adaptive load-shedding	0.0074	0.427	7.16
	0.0338	0.643	4.94
	0.0137	0.820	16.86
Load shedding + modulation control (PMU 6)	0.0087	0.429	8.46
	0.0378	0.647	5.32
	0.0192	0.808	17.33

Table 8.13 Prony analysis fit: frequency measurement at PMU 15 (post-contingency condition)

Control alternative	Amplitude	Frequency (Hz)	Damping (%)
Base case	0.0130	0.394	8.92
	0.0386	0.578	17.97
	0.0059	0.742	13.31
Load shedding + modulation control (PMU 6)	0.0215	0.383	10.48
	0.0641	0.542	28.33
	0.0206	0.706	13.90

References

1. W. A. Mittelstadt (Working group chair), 'Integrated monitor facilities for the western power system: the WECC WAMS in 2003', WECC Disturbance Monitoring Work Group, June 25, 2003.
2. Dmitry Kosterev, Carson W. Taylor, William A. Mittelstadt, 'Model validation for the August 10, 1996 WSCC system outage', *IEEE Transactions on Power Systems*, vol. 14, no. 3, August 1999, pp. 967–979.
3. Eric Allen, Dmitry Kosterev, Pouyan Pourbeik, 'Validation of power system models', 2010 Power and Energy Society General Meeting, July 2010, Minneapolis, MN.
4. *Preliminary Disturbance Report: August 14, 2003 Sequence of Events.* North American Electric Reliability Council, August 15, 2003.
5. Power System Dynamic Performance Committee, Task Force on Identification of Electromechanical Modes, Chair: Juan J. Sánchez Gasca, 'Identification of electromechanical modes in power systems', IEEE/PES Special Publication TP462, June 2012.
6. John F. Hauer, Navin B. Bhatt, Kirit Shah, Sharma Kolluri, 'Performance of "WAMS East" in providing dynamic information for the north east blackout of August 14, 2003', 2004 IEEE Power Engineering Society General Meeting.
7. A. R. Messina, Vijay Vittal, Daniel Ruiz-Vega, G. Enríquez Harper, 'Interpretation and visualization of wide-area PMU measurements using Hilbert analysis', *IEEE Transactions on Power Systems*, vol. 21, no. 4, November 2006, pp. 1763–1771.
8. Enrique Martínez, A. R. Messina, 'Modal analysis of measured inter-area oscillations in the Mexican interconnected system: The July 31, 2008 event', 2011 IEEE Power Engineering Society General Meeting.
9. Arturo R. Messina (ed.), *Inter-Area Oscillations in Power Systems–A Nonlinear and Nonstationary Perspective*, Power Electronics and Power Systems Series, Springer, New York, NY, 2009.

10. Enrique Martínez, 'SIMEFAS: A phasor measurement system for the security and integrity of Mexico's electric power system', 2008 Power and Energy Society General Meeting, July 2008, Pittsburg, PA.
11. Shu-jen S. Tsai, Li Zhang, Arun Y. Phadke, Yilu Liu, Michael R. Ingran, Sandra C. Bell, Dale T. Bradshaw, David Lubkeman, 'Study of global frequency dynamic behavior of large power systems, 2004 IEEE Power Systems Conference and Exposition, October 2004, Phoenix, AZ.
12. A. R. Messina, P. Esquivel, F. Lezama, 'Time-dependent statistical analysis of wide-area time-synchronized data', *Mathematical Problems in Engineering*, vol. 2010, 2010, pp. 1–17.
13. Donald W. Tufts, Ramdas Kumaresan, 'Singular value decomposition and improved frequency estimation using linear prediction', *IEEE Transactions on Acoustic, Speech and Signal Processing*, vol. ASSP-30, no. 4, August 1982, pp. 671–675.
14. E. Barocio, Bikash C. Pal, Nina F. Thornhill, A. R. Messina, 'A dynamic mode decomposition framework for global power system oscillation analysis', accepted for publication in the *IEEE Trans. on Power Systems*, available online: http://ieeexplore.ieee.org/.
15. Prabha Kundur, *Power System Stability and Control*, McGraw-Hill, New York, NY, 2014.

Appendix A

Physical meaning of proper orthogonal modes

A.1 Eigenvalue-based decomposition

An attempt to provide physical insight into the proper orthogonal modes is presented in this appendix.

Following Goebel and Epstein [1], consider a linear system of the form

$$\mathbf{M}\ddot{\mathbf{q}} + \mathbf{K}\mathbf{q} = \mathbf{0} \tag{A.1}$$

where $\mathbf{q}$ is the n-dimensional vector of states, and $\mathbf{M}, \mathbf{K}$ are, respectively, the n-by-n dimensional mass and synchronizing matrices.

The general solution of (A.1) is

$$\begin{aligned} \mathbf{x}(t) &= \sum_{i=1}^{n}(c_i \sin\omega_i t + d_i \cos\omega_i t)\mathbf{u}_i \\ &\sum_{i=1}^{n}(A_i \sin\omega_i t + \varphi_i)\mathbf{u}_i \end{aligned} \tag{A.2}$$

where

$$A_i = \sqrt{c_i^2 + d_i^2}$$

$$\varphi_i = \arctan\left(\frac{d_i}{c_i}\right)$$

and the eigenmodes of the free motion are given by

$$\left(\mathbf{K} - \omega_i^2\mathbf{M}\right)\mathbf{u}_i = \mathbf{0} \tag{A.3}$$

Equation (A.2) can be rewritten in compact form as

$$\mathbf{x}(t) = \sum_{i=1}^{n} a_i(t)\mathbf{u}_i \tag{A.4}$$

where

$$a_i(t) = A_i \sin\omega_i t + \varphi_i$$

Following the general theory in Chapter 3, assume that the displacements $\mathbf{x}(t)$ in (A.4) are used to generate the observation matrix

$$\mathbf{X}(t) = [\,\mathbf{x}_1(t) \quad \mathbf{x}_2(t) \quad \cdots \quad \mathbf{x}_n(t)\,]^T = \begin{bmatrix} x_1(t_1) & x_1(t_2) & \cdots & x_1(t_N) \\ x_2(t_1) & x_2(t_2) & \cdots & x_2(t_N) \\ \vdots & \vdots & \ddots & \vdots \\ x_n(t_1) & x_n(t_2) & \cdots & x_n(t_N) \end{bmatrix} \tag{A.5}$$

where $\mathbf{x}_j(t) = [x_1(t_j), x_2(t_j), \ldots, x_n(t_j)]^T, j = 1, N$.

Define now the matrix of coefficients

$$\mathbf{A}(t) = \begin{bmatrix} \mathbf{a}_1(t) \\ \mathbf{a}_2(t) \\ \vdots \\ \mathbf{a}_n(t) \end{bmatrix}$$
$$= \begin{bmatrix} A_1 \sin(\omega_1 t_1 + \varphi_1) & A_2 \sin(\omega_2 t_1 + \varphi_2) & \cdots & A_n \sin(\omega_n t_1 + \varphi_{n1}) \\ A_1 \sin(\omega_1 t_2 + \varphi_1) & A_2 \sin(\omega_2 t_2 + \varphi_2) & \cdots & A_n \sin(\omega_n t_2 + \varphi_n) \\ \vdots & \vdots & \ddots & \vdots \\ A_1 \sin(\omega_1 t_N + \varphi_1) & A_2 \sin(\omega_2 t_N + \varphi_2) & \cdots & A_n \sin(\omega_n t_N + \varphi_n) \end{bmatrix} \tag{A.6}$$

Substituting (A.6) in the ensemble of data $\mathbf{X}$, gives

$$\mathbf{X} = \mathbf{A}\left[\mathbf{u}_1^T \quad \cdots \quad \mathbf{u}_n^T\right] = \mathbf{a}_1\mathbf{u}_1^T + \ldots + \mathbf{a}_n\mathbf{u}_n^T \tag{A.7}$$

It thus follows that the correlation matrix may be written as

$$\mathbf{C} = \frac{1}{N}\mathbf{X}^T\mathbf{X} = \frac{1}{N}(\mathbf{a}_1\mathbf{u}_1^T + \ldots + \mathbf{a}_n\mathbf{u}_n^T)^T(\mathbf{a}_1\mathbf{u}_1^T + \ldots + \mathbf{a}_n\mathbf{u}_n^T) \tag{A.8}$$

To verify that a modal vector is a POM, we must multiply (A.8) by $\mathbf{u}_j$:

$$\begin{aligned} \hat{\mathbf{C}} = \mathbf{C}\varphi_j &= \frac{1}{N}(\mathbf{X}^T\mathbf{X})\mathbf{u}_j \\ &= \frac{1}{N}(\mathbf{a}_1\mathbf{u}_1^T + \ldots + \mathbf{a}_n\mathbf{u}_n^T)^T(\mathbf{a}_1\mathbf{u}_1^T + \ldots + \mathbf{a}_n\mathbf{u}_n^T)\mathbf{u}_j \end{aligned} \tag{A.9}$$

Further, using the orthogonality characteristics $\mathbf{u}_i^T\mathbf{u}_j = 1$, for $i = j$, (A.9) reduces to

$$\hat{\mathbf{C}} = \frac{1}{N}\left[\mathbf{a}_1^T\mathbf{a}_j\mathbf{u}_1 + \mathbf{a}_2^T\mathbf{a}_j\mathbf{u}_2 + \ldots + \mathbf{a}_j^T\mathbf{a}_j\mathbf{u}_j + \cdots + \mathbf{a}_m^T\mathbf{a}_j\mathbf{u}_m\right] \tag{A.10}$$

in which

$$\mathbf{u}_i \mathbf{a}_i^T \mathbf{a}_j = \begin{cases} \sum_{k=1}^{n}\sum_{m=1}^{n} A_i A_j \sin(\omega_k t + \varphi_k)\sin(\omega_m t + \varphi_m), & i \neq j \\ \frac{N}{2}A_j^2 + \sum_{k=1}^{n}\sum_{m=1}^{n} A_k A_j \cos(\omega_k t + \varphi_k) - \cos(\omega_m t + \varphi_m), & i = j \end{cases}$$

The analysis shows that for a sufficiently large number of snapshots, N, the terms $\mathbf{a}_i^T \mathbf{a}_j \mathbf{u}_i$ vanish, and $\hat{\mathbf{C}} = \mathbf{a}_j^T \mathbf{a}_j \mathbf{u}_j = \alpha \mathbf{u}_j$. In words the eigenvectors (POMs), φ_j, of $\mathbf{C}$, converge to the modal vector $\mathbf{u}_j$; the columns of the left eigenvector are the normalized time modulations $a_i(t) = A_i \sin(\omega_i t - \varphi_i)$ of the eigen modes [2].

A.2 SVD-based POD

This case is treated in [3].

References

1. C. J. Goebel, S. T. Epstein, 'Motion of damped oscillators: Normal modes', *American Journal of Physics*, vol. 48, no. 4, April 1980, pp. 289–291.
2. B. F. Feeny, R. Kappagantu, 'On the physical interpretation of proper orthogonal modes in vibrations', *Journal of Sound and Vibration*, vol. 211, no. 4, 1998, pp. 607–616.
3. J. J. Ayón, E. Barocio, A. R. Messina, 'Blind extraction and characterization of power system oscillatory modes', *Electric Power Systems Research*, vol. 119, 2015, pp. 54–65.

Appendix B
Data for the five-machine test system

B.1 System data

The test system data is given in Tables B.1 through B.4 [1].

Table B.1 Transmission line parameters (in pu on a 100 MVA base)

From bus	To bus	No. of parallel circuits	*R*	*X*	Line charging in MVAR at 1.0 pu voltage
1	2	2	0.08	0.6	2.5
2	4	2	0.08	0.4	10
4	7	2	0.08	0.4	10
4	10	4	0.08	0.4	10
5	7	2	0.08	0.4	10

Table B.2 Transformer parameters (in pu on a 100 MVA base)

From bus	To bus	*R*	*X*	Transformer Tap Ratio[a]
2	3	0.02	0.12	1.02
5	6	0.01	0.06	1.02
10	11	0.02	0.12	1.02
10	12	0.02	0.12	1.02

[a]Tapped side corresponds to the bus listed on the left.

Table B.3 Generator data

Bus	3	6	11	12
X_d	1.920	1.216	1.720	1.670
X_d	1.900	0.756	1.640	1.610
X'_d	0.362	0.336	0.263	0.364
X'_q	1.130	–	0.467	0.536
R_a	0.004	0.002	0.002	0.002
X_L	0.194	0.153	0.145	0.170
T'_{do}	5.690	11.500	4.120	7.800
T'_{qo}	1.500	–	1.475	1.388
H	2.700	7.400	1.600	6.100
D	–	–	–	–
S_1	0.120	0.120	0.120	0.120
S_2	0.480	0.480	0.480	0.480
X_p	0.362	0.336	0.263	0.364
Data MVA base	115	175	115	115

Table B.4 Excitation system data

Bus	3	6	11	12
K_A	400	400	400	400
T_A	0.02	0.05	0.02	0.02
T_B	–	–	–	–
T_C	–	–	–	–
V_{RMAX}	7.30	3.50	7.30	8.20
V_{RMIN}	−7.30	−3.50	−7.30	−8.20
T_E	0.80	0.95	0.80	1.30
K_E	1.0	−0.17	1.00	1.00
S_1	0.05	0.22	0.50	0.50
S_2	0.82	0.95	0.86	1.10
K_F	0.03	0.04	0.03	0.03
T_F	1.00	1.00	1.00	1.00

B.2 Base case load flow condition

The base case load flow condition is given in Table B.5.

Table B.5 Load flow solution: base case condition

Bus	Voltage	Angle (deg)	Load		Generation	
			MW	MV Ar	MW	MVAr
1	1.050	0.000	0.0	0.0	52.69	j0.58
2	1.0313	−8.292	100.0	25.00	0.0	0.0
3	1.050	−2.048	0.0	0.0	100.0	22.68
4	1.0260	−13.774	100.0	25.00	0.0	0.0
5	1.0649	1.602	0.0	0.0	0.0	0.0
6	1.050	4.437	0.0	0.0	090.0	−2.19
7	1.0360	−7.612	030.0	15.00	0.0	0.0
10	1.0254	−14.000	200.0	50.00	0.0	0.0
11	1.050	−7.771	0.0	0.0	100.0	26.82
12	1.050	−7.771	0.0	0.0	100.0	22.68

Reference

1. *A Study of Static Reactive Power Compensators for High-Voltage Power Systems*, Prepared by Advanced Systems Technology Division and Transmission and Distribution Systems Engineering Department, Westinghouse Electric Corporation, Contract 4-L60-6964P, Final Report, May 1981.

Appendix C
Masking techniques to improve empirical mode decomposition

This appendix discusses extensions to conventional empirical mode decomposition (EMD) analysis to study oscillatory dynamics.

C.1 Energy-based masking technique

In [1], EMD with masking technique was introduced to address the problem of mode mixing. In [2–4], a systematic procedure for constructing the masking signals is proposed.

The key idea in these procedures is to insert a masking signal to prevent lower frequency components from being included in the intrinsic mode function (IMF). Among the various approaches proposed, methods based on the EMD itself (energy information) are of particular interest since they do not rely on any external information [4].

Assume, in order to introduce these ideas, that the EMD procedure is applied once, and let $A_1(t)$ and $f_1(t)$ be the instantaneous amplitude and frequency respectively of the first, raw IMF. As noted in [1], the first IMF is expected to contain the highest frequency component of the signal.

While this idea can be applied recursively, this issue has been barely addressed in power system literature.

Masking signal EMD method

1. Perform EMD on the original signal $x(t)$. Use only the first IMF, $c_1(t)$, which is expected to contain the highest frequency component of the signal, $f_{max}(t)$. Obtain $A_1(t)$ and $f_1(t)$ using Hilbert analysis (or any other approach that computes the instantaneous amplitude and frequency).
2. Compute the energy weighted mean of $f_1(t)$ over L samples using the energy-weighted instantaneous frequency

$$\bar{f}_1(t) = \frac{\sum_{i=1}^{L} A_1(i) f_1^2(i)}{\sum_{i=1}^{L} A_1(i) f_1(i)}$$

(*Continues*)

(*Continued*)

Masking signal EMD method
3. Construct the masking signal $mask_1(t) = M_1 \sin(2\pi(m\bar{f}_1)t)$ where $$M_1 = \max_{i=1,\ldots,L} A_i(i) \quad \text{and} \quad m > 1$$
4. Perform EMD on $x_+(t) = x(t) + mask_1(t)$ and $x_-(t) = x(t) - mask_1(t)$. Obtain the IMFs for $c_{i_+}(t)$ and $c_{i_-}(t)$, $i = 1,\ldots, n$, and the residues $r_{n_+}(t), r_{n_-}(t)$. The IMFs and residues of the signal are then given by $$c_i(t) = \left(\frac{c_{i_+} + c_{i_-}}{2}\right), \quad i = 1, 2, \ldots, n$$ $$r_n(t) = \left(\frac{r_{n_+} + r_{n_-}}{2}\right)$$
5. Use the next masking signal to perform steps 2–4 iteratively, using each masking signal while replacing $x(t)$ with the residue obtained at each iteration until $n - 1$ IMFs containing the frequency components $f_2, f_3, \ldots, f_n$ are extracted.

The masking signal EMD method can be summarized as follows:

Refer to [2–4] for extensions of this idea and a detailed explanation of numerical algorithms to choose the masking signals. The total effect of these three operations is to separate the low-frequency components from the high-frequency modes.

It should be noted that the selection of the masking signal is not unique. Moreover, the choice of the signal amplitude A_o can affect the performance of the algorithm, although numerical experience suggests that reasonable results can be obtained using values which are not much larger than the highest frequency.

References

1. R. Deering, J. F. Kaiser, 'The use of a masking signal to improve empirical mode decomposition', *Proceedings of the IEEE International Conference on Acoustics, Speech and Signal Processing* (ICASSP '05), vol. 4, 2005, pp. 485–488.
2. N. Senroy, S. Suryanarayanan, 'Two techniques to enhance empirical mode decomposition for power quality applications', *Proceedings of the IEEE Power Engineering Society General Meeting*, Tampa, FL, 2007, pp. 1–6.
3. N. Senroy, S. Suryanarayanan, P. F. Ribeiro, 'An improved Hilbert–Huang method for analysis of time-varying waveforms in power quality, *IEEE Transactions on Power Systems*, vol. 22, 2007, pp. 1843–1850.
4. D. S. Laila, A. R. Messina, B. C. Pal, 'A refined Hilbert–Huang transform with applications to inter-area oscillation monitoring', *IEEE Transactions on Power Systems*, vol. 24, no. 2, 2009, pp. 610–620.

Index

Note: Page numbers followed by '*f*' and '*t*' indicate figures and tables respectively.

architecture, WAMS
- centralized 11–12, 12*f*
- data fusion model
  - data assimilation 20–2
  - elementary data fusion strategy 19*f*
  - hierarchical multiblock data models 19–20
  - intelligent synchrophasor 14–16
  - multiarea power system 16, 17*f*
  - multiple data sets correlation 16
  - spatio-temporal information 18
- data recording system 10*f*
- feature extraction/pattern recognition 10*f*
- hierarchical/distributed 12–13, 13*f*
- hybrid WAMS architectures 13
- multiblock and single-block models 22–3

auto-associative neural networks (AANN), nonlinear PCA 112–13

binary connectivity matrix 39

blind source separation (BSS)
- complex BSS formulations 121–3
- lagged variables 120–1
- spatial amplitude function 123
- spatial phase function 123
- temporal amplitude function 123
- temporal phase function 123–4
- visualize system behavior 216–17

data analysis
- damping estimation 197, 199*t*
- instantaneous damping 201–5
- instantaneous energy 200–1
- instantaneous frequency 200
- instantaneous parameters 197, 199*f*
- mode shape characterization 196–7
- multimodal data 209
- multitemporal, multiscale analysis 205–9
- performance evaluation 210–11

data fusion
- data assimilation 20–2
- data oriented (low-level fusion) 15
- elementary data fusion strategy 19*f*
- feature-level fusion 15, 22*f*, 23*f*
- hierarchical multiblock data models 19–20
- high-level fusion 15
- intelligent synchrophasor 14–16
- multiarea power system 16, 17*f*
- multiple data sets correlation 16
- principles 101–2
- processing chain 102*f*
- spatio-temporal information 18
- task oriented (feature extraction) 15

data pre-processing and transformation
- bandpass filtering and denoising 104–5
- damage identification 134
- local-level fusion 105

data processing and feature extraction
- ambient stimulus, response under
  - ensemble system response 90
  - formulation of model 87–8
  - modal response 89

application to measured data
HHT analysis 92–4
wavelet analysis 94–6
mutivariate multiscale analysis
Koopman analysis 83–7, 84*f*, 85*f*
multi-signal prony analysis 82–3, 83*f*
power oscillation monitoring 64–5, 65*t*
time-frequency representations
dynamic harmonic regression (DHR) 76–81
Hilbert-Huang analysis 65–72
Teager-Kaiser operator 75–6
wavelet analysis 72–5
disturbance and anomaly detection 132–4
dynamic harmonic regression (DHR)
forecasting 81
Kalman filter and smoothing algorithms 78
state space modeling framework 77–8
time-variable parameters 78–9
trend extraction 79–81

eigenvalue-based decomposition 227–9
empirical mode decomposition (EMD) 67–70
empirical orthogonal function (EOF) analysis 29–33
energy 138–41
energy-based masking technique 235–6
entropy 138–41
entropy-based power monitoring 141–2

five-machine test system
base case load flow condition 233*t*
excitation system data 232*t*
generator data 232*t*
transformer parameters 231*t*
transmission line parameters 231*t*

hierarchical multiblock data models 19–20
high-dimensional pattern recognition-based monitoring
data clustering
computational issues 148
k-nearest neighbors 147–8
hybrid schemes 150
numerical example 148–50, 149*f*, 149*t*
sparse diffusion implementation 146–7
Hilbert transform, near real-time analysis and monitoring 169–73
Hilbert-Huang transform (HHT) analysis
analysis procedure 93
damping and frequency characterization 70–1
empirical mode decomposition 67–70
IMFs and spectra 92*f*
nonlinear and nonstationary time series 71–2
phase characterization 71
PMUs 2, 3, and 6 94*f*

intelligent synchrophasor data fusion 14–16, 14*f*, 15*f*
intrinsic mode functions (IMFs) 66

Koopman mode analysis
mutivariate multiscale analysis 83–7, 84*f*, 85*f*
near real-time analysis and monitoring 181

monitoring, power system. *see* power system health monitoring
Moran coefficient *I*(x) 40
multi-signal prony analysis 82–3, 83*f*
multi-variate spatio-temporal process
EOF analysis 29–33
mean value of time series 36–7
SVD-based proper orthogonal decomposition 33–6

multiblock POD analysis
  multiscale PCA 117–18, 117*t*
  partial least squares (PLS) 118–19
  raw-level data 114
  sensor level 114–16, 115*f*, 116*f*
  temporal scales 116–17
multisensor multitemporal data fusion
  architecture 102–3
  blind source separation (BSS)
    complex BSS formulations 121–3
    lagged variables 120–1
    spatial amplitude function 123
    spatial phase function 123
    temporal amplitude function 123
    temporal phase function 123–4
  data compression 106–9
  data pre-processing and transformation
    bandpass filtering and denoising 104–5
    local-level fusion 105
  decision level 104
  elementary data fusion 110*f*
  feature selection 106
  feature-extraction level 103, 105–6
  feature-level fusion 104
  filtering and multiscale monitoring 109–11
  individual scales 109
  Koopman mode analysis 126*f*, 128
  multiblock POD (PCA) analysis
    multiscale PCA 117–18, 117*t*
    partial least squares (PLS) 118–19
    raw-level data 114
    sensor level 114–16, 115*f*, 116*f*
    temporal scales 116–17
  nonlinear PCA 112–13, 119
  POD method 124, 126
  single-scale PCA method 112

near real-time analysis and monitoring
  abnormal operation detection
    Hilbert transform 169–73
    local mean speed 173–5, 176*f*
    nonlinear and/or nonstationary signal processing 166–9
  damage and disturbance detection
    event trigger 164
    linear filtering 164–6, 164*f*
    recorded test signal 166*f*
  data processing and conditioning
    EMD-based filtering 162–3
    wavelet denoising and filtering 160–2
  pattern recognition-based disturbance detection 176–7
  recursive processing methods
    linear regression 182–3
    system oscillatory modes 183–8
  sliding window-based methods
    conventional HHT analysis 177–9
    Koopman mode analysis 181
    numerical example 180–1
nonlinear spectral dimensionality reduction
  diffusion maps 44–6
  grouping trajectories 47
  overview of/characteristics 44*t*
  time series interpretation 46–7

oscillation loss analysis
  AR spectra of frequency measurements 195*f*
  frequency transients 194*f*
  operational context 192
  PMU measurement locations 194*t*
  time windows 194

partial least squares (PLS)
  data assimilation 20*f*
  multiblock POD (PCA) analysis 118–19
pattern recognition analysis
  diffusion map analysis 211–14
  isomap method 215
  Laplacian eigenmap analysis 215, 215*f*

PCA analysis, voltage and reactive power monitoring 155
phasor data concentrators (PDCs) 10, 12, 13*f*
phasor measurement units (PMUs) 10, 13*f*
power oscillation monitoring 64–5, 65*t*
power system health monitoring
  disturbance and anomaly detection 132–4
  high-dimensional pattern recognition-based monitoring
    data clustering 147–8
    hybrid schemes 150
    numerical example 148–50
    sparse diffusion implementation 146–7
  modal-based methods
    entropy and energy 138–41
    entropy-based detection 141–2
    filtering and data conditioning 134–8
  real-time spatio-temporal databases 132
  voltage and reactive power monitoring
    measured data 150–1
    POD/PCA analysis 155
    statistical approach 151–5
  wide-area inter-area oscillation monitoring 143–6
proper orthogonal decomposition (POD) analysis
  multisensor multitemporal data fusion 124, 126
  visualize system behavior 216–17
  voltage and reactive power monitoring 155

real-time spatio-temporal databases 132

sensor placement, spatio-temporal modeling
  constrained sensor placement 54–8
  problem formulation 53–4
sifting 66
single-scale PCA method 112
singular value decomposition (SVD) analysis 33–6
spatial amplitude function 123
spatial phase function 123
spatio-temporal modeling
  data fusion 18
  dimensionality reduction
    nonlinear spectral dimensionality reduction 43–7, 44*t*
    proximity (similarity) measures 42–3, 43*f*
  large system response 48
  multivariate processes
    EOF analysis 29–33
    mean value of time series 36–7
    SVD-based proper orthogonal decomposition 33–6
  sensor placement
    constrained sensor placement 54–8
    problem formulation 53–4
  small-signal response 48
  spatial interpolation
    derivation of weights 40–1
    practical issues 41
    problem of monitoring 37*f*, 38
    similarity measures 38–9
    spatial structures 40
  statistical analysis 49–51
  ten-bus, 5-machine test system 49–51, 52*f*
  visualization of data 28–9, 28*f*
SVD-based proper orthogonal decomposition 33–6

Teager-Kaiser operator 75–6
temporal amplitude function 123
temporal phase function 123–4

validation of power system model
  control performance evaluation 221–4, 225*f*
  large system performance 218, 220*f*
  small signal performance 218

visualize system behavior
data analysis
damping estimation 197, 199*t*
instantaneous damping 201–5
instantaneous energy 200–1
instantaneous frequency 200
instantaneous parameters 197, 199*f*
mode shape characterization 196–7
multimodal data 209
multitemporal, multiscale analysis 205–9
performance evaluation 210–11
oscillation loss analysis
AR spectra of frequency measurements 195*f*
frequency transients 194*f*
operational context 192
PMU measurement locations 194*t*
time windows 194
pattern recognition analysis
diffusion map analysis 211–14
isomap method 215
Laplacian eigenmap analysis 215, 215*f*
POD/BSS analysis 216–17
validation of power system model
control performance evaluation 221–4, 225*f*
large system performance 218, 220*f*
small signal performance 218
voltage and reactive power monitoring
measured data 150–1, 151*f*
POD/PCA analysis 155
statistical approach 151–5, 152–3*f*, 154*t*

wavelet analysis
application to measured data 94–6
near real-time analysis and monitoring 160–2
phase difference 73–4
wavelet transform 72
with EMD 74–5
wide-area inter-area oscillation monitoring 143–6
wide-area monitoring systems (WAMS)
advanced sensing and metering 5
applications 3, 11
bandwidth requirements 5
components 2*f*
data collection and management 3–4
data paucity 4
incomplete data analysis 5
sensing techniques 4–5
sensor placement 5
sensor selectivity and data fusion 4
structure 2, 10, 10*f*